# Chemically Imbalanced

# Chemically Imbalanced

Everyday Suffering, Medication, and
Our Troubled Quest for Self-Mastery

JOSEPH E. DAVIS

The University of Chicago Press
Chicago and London

The University of Chicago Press, Chicago 60637
The University of Chicago Press, Ltd., London
© 2020 by The University of Chicago
All rights reserved. No part of this book may be used or reproduced in any
manner whatsoever without written permission, except in the case of brief
quotations in critical articles and reviews. For more information, contact the
University of Chicago Press, 1427 E. 60th St., Chicago, IL 60637.
Published 2020
Printed in the United States of America

29  28  27  26  25  24  23  22  21  20      1 2 3 4 5

ISBN-13: 978-0-226-68654-7 (cloth)
ISBN-13: 978-0-226-68668-4 (paper)
ISBN-13: 978-0-226-68671-4 (e-book)
DOI: https://doi.org/10.7208/chicago/9780226686714.001.0001

Library of Congress Cataloging-in-Publication Data

Names: Davis, Joseph E., author.
Title: Chemically imbalanced: everyday suffering, medication,
    and our troubled quest for self-mastery / Joseph E. Davis.
Description: Chicago : University of Chicago Press, 2020. | Includes
    bibliographical references and index.
Identifiers: LCCN 2019025708 | ISBN 9780226686547 (cloth) |
    ISBN 9780226686684 (paperback) | ISBN 9780226686714 (ebook)
Subjects: LCSH: Affective disorders—Treatment. | Neurobehavioral
    disorders—Treatment. | Neurobiology.
Classification: LCC RC537 .D35 2020 | DDC 616.85/27—dc23
LC record available at https://lccn.loc.gov/2019025708

♾ This paper meets the requirements of ANSI/NISO Z39.48-1992
(Permanence of Paper).

*For Monica*

We must rid ourselves of the delusion that it is the major events which have the most decisive influences on us. We are much more deeply and continuously influenced by the tiny catastrophes that make up daily life.

Siegfried Kracauer, *The Mass Ornament*

It is not to the different that one should look for understanding our differentness, but to the ordinary.

Erving Goffman, *Stigma*

CONTENTS

We have all heard the story. The brain is the last scientific frontier and the unraveling of its mysteries is playing an increasingly central role in how we understand the world and ourselves. Breathless reports in the popular press and in the best-selling writing of scientists inform us that we are in the midst of a revolution, entering a new and enlightened era in which many of our most persistent human problems will be conquered. Biological explanations of mental life are sweeping away long-standing philosophical problems—mind-body and nature-nurture—and the vexing enigmas of human subjectivity and consciousness. Significant advances in genetics, biochemistry, and neuroscience are yielding breakthroughs in the understanding of neural mechanisms and the physiology of human thought, emotion, and behavior. Psychiatry, breaking free of its psychological past, is becoming "clinical neuroscience" and will soon transform the way it treats mental disorders. The days of the old "folk psychology" and such long outdated notions as the soul are finally coming to an end.

That is the story, and judging from the book sales figures, positive media coverage, and other evidence over the past several decades, neurobiological accounts of mind, self, and behavior have been eagerly embraced by the general public. Why? The enthusiastic reception, it seems safe to say, is not in response to the discrediting of the old philosophies, or the appearance of new treatments, or the scientific discovery of new phenomena. Though there are countless new insights, the reality is far more pedestrian than the hype. There is little settled knowledge of disorders or treatments or the relationship of mind to body. In fact, many of the claims about the relation of mind and mental states to brain are not really scientific at all and cannot themselves be tested in any empirical way. They rest not so much on a theory as on changed assumptions about human being. While the promise of neuroscience responds to a widespread yearning for concreteness and a promise of unambiguous solutions to intractable problems, the explanatory force of its insights and their actual productivity is not nearly enough to explain the public appeal.

Something else, something in our common culture, is afoot. This book is a field report on how ordinary people dealing with painful everyday struggles with loss and failure and limitation engage with the new and psychologically depthless talk of neurobiology. These encounters, in turn, serve as a kind of stethoscope on our underlying condition, on a change in the way that we imagine ourselves and how to get on in our world.

Like many of her college peers, Kristin, twenty-one, had experimented with Adderall. Her first encounter with the preferred medication for treating attention deficit/hyperactivity disorder (ADHD) came at a party, where she met a student who was "just, like, crazy," animated and making people laugh. Not long into their conversation, he told Kristin that he was taking Adderall. Getting started on it had been easy, he explained, because his mother liked "pathologizing everything" and had sent him to a psychiatrist to get an evaluation for his trouble focusing in high school. After a brief session—"like two minutes," the young man said—the psychiatrist wrote a prescription. Impressed by his story, Kristin left the conversation thinking they both had the same sort of thing. She wanted to try the drug.[1]

She didn't have to wait. A friend at the party gave her a pill, and soon she was feeling so good that she worried she might be experiencing an "artificial high." She quickly got over her reservations, however, and, supplied by friends, she began taking the medication on a regular basis. She grew convinced that the drug could improve her social life and her ability to "get things done."

Kristin believed she needed help on both fronts. From a relatively affluent home, she had done very well in high school and then matriculated at a selective college known for its academic rigor. She anticipated challenging classes and hoped for a vibrant social life. But after two years, she was disappointed and frustrated. The other students were just like Kristin herself—"kind of shy, artistic, very smart, but [they had] a hard time interacting." She found them unhappy, and the social environment stifling. She frequently felt sad and decided to transfer to a college with more of a party-school reputation.

Kristin's transition to the new environment was rocky: Once classes began, her expectations of an active social life quickly gave way to a pervasive sense of inadequacy. She was constantly angry with herself and, perhaps even worse, afraid to take the initiative to meet new people. Not long into the new semester, she decided to seek help.

At the student health center, she met with a counselor and found the experience encouraging and helpful. It "was like going to confession," she said, "only more confessional." She let the secrets she had been keeping inside "just spill out" and felt relief in doing so. But after three sessions, she had had enough. Though still "afraid of things inside," she saw no reason to go back. She thought the counseling would be too much of a "hassle" and unlikely to achieve anything "very lofty."

Kristin wanted something else. She thought that many of the problems in her social life reflected a larger issue of "focusing." Her parents had often told her that she needed more discipline. She had assumed they were right, not because her grades were poor or she had trouble completing specific tasks but because she believed she wasn't living up to her full potential. She had allowed herself to dabble instead of attempting to concentrate on one particular thing. "I want to do everything in the world and hear everything and do—like, just be a star," she said, even while ruefully acknowledging that she wasn't particularly interested in her studies or involved in many extracurricular activities. Then came her exhilarating experience with the fellow at the party and trying Adderall.

Kristin went to the college learning center to pursue an ADHD diagnosis, but the specialist suggested that she read some books on effective learning strategies. Undeterred, she decided to seek the help of a psychiatrist in private practice. What she encountered could not have been more different from her sessions with the counselor at student health. The counselor wanted to know what was going on with her life. Instead of "talking it out," the psychiatrist, by contrast, wanted to catalog her symptoms and did not ask about her "personal perspective on them" or how she "might focus on addressing them." The psychiatrist's goal was a diagnosis, and according to Kristin, she told her that she "had some of the symptoms, not all of them, of ADHD, and that a stimulant might help." Then she wrote Kristin a prescription. The session took about an hour and a half.

Kristin embraces the diagnosis, though on her own terms. ADHD is classified as a mental disorder in the *Diagnostic and Statistical Manual of Mental Disorders* (DSM), the American Psychiatric Association's resource book for psychiatric diagnoses. But Kristin doesn't see it that way and even contrasts her definition with that of professionals. For her, conditions like ADHD can be interpreted in different ways, and this very flexibility is part of what makes the diagnosis fit. In her creative activities, she sees distractibility as conducive to innovation, while a debility in other areas requiring sustained concentration on one thing. Having benefits as well as drawbacks makes it different from a mental disorder. With schizophrenia, she says, "your brain just takes over and you can't get over it."

Though Kristin does not see her ADHD as a mental disorder, she does believe it is caused in significant part by a genetic aberration. The psychiatrist did not offer this explanation, but she has talked with friends and read about ADHD. She cites the experience of family members as contributing to this view. Her grandmother and both of her siblings, she notes, have a hard time focusing on things, especially her older sister, who urged her to seek a prescription and who now believes she too has ADHD. She also finds some confirmation of the biological interpretation in the fact that the drug seems to facilitate her classwork.

For Kristin, her ADHD diagnosis helps fill the explanatory gap between her self-understanding and her circumstances. It confirms that her struggles are both real and legitimate: "actual problems" that arise from external causes, in her biology. These problems are things that you can address. They are unlike "abstract moral failings," Kristin's term for speaking about fixed dispositions in the self or personality, about something you *are*, such as a "bad person." But with the neurobiological account, she can stop blaming herself for "being lazy" or a "poor listener." No such moral or mental framework is relevant.

While Kristin credits the stimulant with some positive results, it has not made the difference she hoped it would. "I guess I used to go through a day," she says, "and imagine that I had only worked 50 percent as hard as I could have and that one of these days, I was going to do 100 percent and I was just going to be, it was just going to be great." On the medication, she does push herself somewhat harder, and she finds that the drug makes her feel smarter and more interested in her class work. But there has been no dramatic improvement, and she also finds it hard to relax when taking it. Sometimes she feels that the medication improves her social interactions, making her both more interesting and more interested in those around her. But at other times she worries that, socially, she is "just kind of a pain." The drug is helpful, yes, she concludes, but "not like a wonder drug."

---

Stories like Kristin's are not unusual.[2] They are only too common. Tens of millions of people, dealing with everyday struggles, have been diagnosed with a mental disorder and are being treated with a psychotropic medication.[3] In order to better understand this remarkable and highly publicized phenomenon, I and research assistants interviewed eighty people, adults ages eighteen to sixty-three.[4] These participants responded to an advertisement I ran in the Chicago, Baltimore, and Boston metropolitan areas, as well as in a part of central Virginia that includes two small cities, Charlottesville and Harrisonburg. The ad asked potential volunteers if they struggled with being sad, with being anxious in social situations, or with concentration and attention problems, and if they would be willing to talk about their experience. I chose these three types of struggles because they are among the most widely reported forms of distress in America today.[5] I worded the ad the way I did in order to recruit people like Kristin, with a diagnosis and taking medication, as well as people who were coping in other and nonmedical ways. By comparing different explanatory accounts and treatment strategies, I sought to better understand what was unique to each.

For a generation, debate has waxed and waned and waxed again over the expanding number of people who are diagnosed with a mental disorder and treated by means of prescription drugs. One side, the psychiatric, sees progress. This side, the medical mainstream, asserts that the growing incidence of diagnosis and treatment of emotional, behavioral, and cognitive problems simply reflects the high numbers of individuals with debilitating mental disorders. The American Psychiatric Association, for instance, claims that one in four Americans experiences a mental disorder in any given year. In this view, people with pathological conditions are finally getting the help they need. All the diagnosing and prescribing are for people who are mentally ill. The same idea comes across in the popular press and in some of the best-selling books by sufferers, who write of their condition from a psychiatric perspective.

But while mental illness is a significant public and clinical issue, most of those diagnosed with emotional and behavioral problems do not fall into this category.[6] We know that the DSM diagnostic thresholds for many listed disorders, such as depression, social anxiety disorder, and ADHD, are low, open-ended, and fail to take account of the context of experience.[7] Before being redefined in the 1960s and 1970s, depression was a relatively rare diagnosis.[8] Social anxiety disorder, which made its first appearance in the DSM in 1980, was virtually unknown before the late 1990s, when suddenly estimates appeared of a hidden population of ten to sixteen million sufferers, making it the third most common mental disorder in America.[9] Some one in five American boys of high school age have been diagnosed with ADHD, a proportion far exceeding even the highest epidemiological estimates.[10] We know that many professionals are making diagnoses based on reports of distress without regard for even the broad DSM diagnostic criteria, are spending little time interacting with patients, and are making minimal efforts to explore the circumstances of patient problems.[11] The most comprehensive epidemiological study of mental health to date found that in only half the treated cases of "mental disorder" did the person being diagnosed meet even the minimal diagnostic criteria.[12] And we know that people now commonly appropriate a diagnosis, adopting medical language for their problems prior to physician confirmation or even contrary to it.[13]

What about these people? People dealing not with serious mental illness but with the sort of emotional distress, trying circumstances, and behavioral troubles that practically anyone might experience at some point or other? These were the people I wanted to talk with, the people who account for the bulk of the diagnoses and prescriptions, whose experience cannot (unproblematically) be contained under the rubric of mental disorder. The DSM itself says that expectable responses to a common stressor or loss, socially deviant behavior, and conflicts that are primarily between the individual and society are not mental disorders.[14] Yet for the people I interviewed, these are the very kinds of troubles of which they speak.[15] How, I inquired, do they make sense of their experience, their "everyday suffering"?

The issue of improper diagnosis brings us to the other side in the debate over the large numbers of those diagnosed and treated with drugs. Many books and articles have been written challenging the psychiatric perspective and the medicalization and widespread treatment of common personality traits and forms of distress as mental disorders.[16] Here are three popular examples, written by a literary critic, a sociologist, and a psychiatrist, respectively:

> *Shyness: How Normal Behavior Became a Sickness*
> *The Medicalization of Society: On the Transformation of Human Conditions into Treatable Disorders*
> *Saving Normal: An Insider's Revolt against Out-of-Control Psychiatric Diagnosis, DSM-5, Big Pharma, and the Medicalization of Ordinary Life*[17]

These books and others like them tell a valuable story about the proliferation of DSM disorder categories, aggressive marketing by the pharmaceutical companies, and, even more ominous, these companies' efforts to influence scientific and clinical research.

The shortcoming of these books is that they typically fail to attend to the suffering of those nondisordered people who turn to doctors for help. They concentrate, instead, on the actions of professionals and industry and on larger institutional pressures and changes that contribute to the re-definition of "human conditions" or "normal behavior" in medical terms. They usually stop there, however, without exploring how these patients are actually struggling.[18] They characterize the concerns at stake for people in minimalist terms, as "vague complaints" or "ordinary unhappiness." Or they imply that dissatisfaction with normal life is being directly gener-ated by the promotion of disorder categories ("disease mongering," "sell-ing sickness") and intervention options.[19] There is a common tendency, for instance, to treat pharmaceutical advertising as a highly deceptive prac-tice, fooling people into pursuing medications they don't actually need. Little is said about why individuals' experience might stand out to them as distinctive and in need of intervention and what suffering, if any, might be involved.[20] Little is said about *their* reasons for seeking medicalized ap-proaches and what effect these approaches are having on them and their understanding of themselves.

In listening to their stories, I sought to better understand their needs and purposes—to hear what they were saying, not just in their words but in what might stand in the background informing them. How, for instance, I inquired, are medical symbolism, neurobiological concepts, and pharma-ceuticals being taken up as social objects or narratives that people invoke to make sense of and alter their life trajectories? Despite what one hears, the heavy use of prescription psychotropic medications is not a recent phe-nomenon. It goes all the way back to the 1950s, and by the late 1960s, adults were using psychotropics at rates comparable to those of today. The crucial change, I found, is something different, a shift much quieter and less measurable than drug use. It is a change in "imaginary," in the terms in which ordinary people imagine self and suffering, its causes and its resolution. This change, the subject of this book, is the master trend for which I seek a "thick" (theoretical, historical, contextual) understanding and assessment.[21]

Before turning to the new imaginary, I need to spell out in greater de-tail what I mean by "everyday suffering" and the "predicaments" that it places people in. I need concepts like these, rather than psychiatric terms like "mental disorder," to make sense of what participants said about their experience and how it figured in their interpretations.[22] In the language of psychiatry, much of what is important here, including central questions of self and social norms, are simply invisible. Psychiatric language excludes them a priori.

## Suffering in Everyday Life

With "everyday suffering," I want to introduce a concept that unlike, say, "condition," always references the lived experience of the sufferer and does not imply any pathology. It names distress and struggle, the painful reality they constitute, and the strong desire to be free of them. And it specifies that this is a type of suffering that arises in everyday life. The "everyday," by dictionary definition, means routine or unremarkable. But in the social sciences, including sociology and anthropology, and in political and historical writing "from below," the "everyday" refers to the practical and felt experience of ordinary people, and efforts to understand it are set against abstract, formal-institutional, and elite-focused accounts of social life. Everyday life is the ongoing arena of practical activity and relationships guided by implicit, "common-sense" knowledge of the world.[23] It is the setting in which selfhood is constituted.[24]

Everyday life, as we all know from experience, includes struggle. Especially under conditions of rapid social change, such as in our own time, it is subject to disruptions and setbacks of varying duration and intensity—divorces, job losses, moves, children's struggles, moral dilemmas, crises of faith, illnesses, deaths of loved ones, and much more. And no list could capture the possible failures, rejections, humiliations, and other troubles that one might have to endure. Difficult and often highly charged experiences, like those reported in this book, interfere with the normal routines of life and can lead to suffering. Granted, "suffering" is a formidable word, and we often prefer to limit its application to accounts of intense physical pain or devastating experiences of injury, deprivation, or trauma. None of those whom I interviewed spoke of their experience as "suffering." Yet, I could not find a better word. And it is not uncommon to speak of suffering in broader terms. Commenting on the suffering literature, sociologist Iain Wilkinson notes some everyday emotional and behavioral ordeals. "We identify suffering," he writes, "as taking place in experiences of bereavement and loss, social isolation, and personal estrangement." It can "comprise feelings of depression, anxiety, guilt, humiliation, boredom, and distress."[25]

Along with emotional distress, everyday suffering includes affronts to the self and a sense of abnormality. While the capacity of suffering to disrupt and confound creates many kinds of challenges, tribulations of the self, experiences that threaten our self-understanding and social standing, are a defining feature.[26] Recall Kristin's story. The standard she evaluates herself against requires one to stand out from the crowd and manifest one's exceptionalness in the attention received from others and in one's engaging, witty, and self-assured mode of interacting.[27] For Kristin, falling short of this standard was disheartening and puzzling, especially when transferring to the new college brought no improvement. At that point, she had no alternative explanation or idea of what to do, and she sought professional help.

Kristin's suffering made no sense to her. She had accepted her parents' idea that she needed more discipline, but somehow working harder didn't address what was actually troubling her. Her problem lay elsewhere, not so much in failing to achieve particular goals but in the *self* that failed to stand out and achieve more. In a discussion of the experience of despair, the nineteenth-century Danish philosopher Søren Kierkegaard illustrates the distinction, observing that "an individual in despair despairs over *something*." But in "despairing over *something*," such as not achieving his ambitions, "he really despaired over *himself* . . . and cannot bear to be himself."[28] While despair seems too strong a word for Kristin's experience, accounting for her inability to meet the standard, a standard or norm of being, confronted her with a very basic and prior interpretive question about herself, namely, did her experience indicate that there was something wrong with *her*, with who she *was*?[29] Some version of this question was asked by nearly everyone interviewed for this book.

This dilemma arises because the very nature of the suffering touches on how we think about ourselves, our purposes, and our place in the world. Consider a comparison case. Public campaigns to destigmatize mental disorders often invoke an analogy with chronic medical conditions like diabetes. The analogy has many problems, including the fact that medical conditions such as diabetes are not fundamentally about self-understanding and aspiration. While having such a condition might lead to everyday suffering (the distress of dealing with a chronic condition with life-and-death implications), the condition itself is defined, diagnosed, and treated from the outside, so to speak, in terms of physiological function. The suffering discussed here is different. While these experiences—changes in feelings, perceptions, thoughts, and actions—have physiological effects, their meaning is directly bound up with our status as persons and our relations with others. There is no blood test for such suffering. Nor could there be. What makes experience feel the way it does, or what causes it to induce shame, or worry, or disappointment, is not independent of the person who has the experience. It reflects a first-person evaluative stance, to which there can be no "objective" or dispassionate access from the outside.

The evaluative in everyday suffering necessarily involves an assessment of our experience against what we take to be the way things are. That something is wrong can be known or felt only by reference to a social standard or ideal, however tacit—what is normal, expected, appropriate, praiseworthy—a standard or ideal that both matters to us and stands over us.[30] Our specific judgments are structured by the languages and assumptions of the ongoing community of which we are a part and include comparisons, whether articulated or not, to how well we think other people meet the norms. For Kristin and others, subjective comparisons of this kind indicated that there was something different about them, something confusing and disturbing.

Suffering, from the Latin, *suffero*, "to undergo," is something that happens *to* us. Even in cases in which our own actions led to trouble, the suffering as such was not something we intended to bring about. In suffering, something valuable to us is jeopardized, and part of what makes this experience so threatening and distressing is that it defies our intentions. The experience is *against* us. It creates confusion and disorientation; it is an experience of frustrated becoming that is hard to understand or express, both unsettling and isolating. For the interview participants, why they struggled as they did was not obvious to them. To varying degrees they felt at sea, aware of their inability to act effectively so as to overcome or cope with whatever was causing them to suffer. This lack of control intensified their suffering and brought an insistent demand for an explanation and a response.[31]

## Predicaments

Everyday suffering places people in what I will call a predicament. I use the term in the sense given by the political theorist William Connolly: "A predicament is a situation lived and felt from the inside. It is something you seek to ameliorate or rise above."[32] The predicament is the quandary of how to locate events and experiences in an explanatory context or account that can provide understanding and guide efforts at relief or escape (which might simply mean continuing on in the best way one knows how).[33] In the interviews, the predicament took the general form of "Why can't I?" questions with respect to specific standards. Like Kristin, many people expressed challenges in terms of social performance. They described difficulties in group situations, such as at parties, when meeting strangers, or when speaking in front of others. Their quandary might be expressed by the question, "Why can't I carry off social encounters with the ease and confidence that other people do?" A related group expressed their dilemma in terms of underperformance in work or school contexts. For those in school, grades and organizational challenges were the great concerns, while for the other adults, work failures and limits on career opportunities loomed largest. The key question came down to this: "Why can't I work as well as I'm supposed to?"

In another common predicament, participants spoke of their suffering in terms of a loss or serious disruption in significant relationships. Through some event, such as a death, divorce, or romantic breakup, they had lost, or lost contact with, one or more people close to them (friend, spouse, parent, etc.). They did not define their suffering in terms of the loss as such (though a few did emphasize continuing to feel the loss of companionship). Rather, the main question they asked might be summarized as follows: "At this point, after the loss or change, why am I still being hampered or held back, unable to move on?" They don't regard their emotions as sufficiently under their control, or they are feeling a continuing emptiness that should be filled by now.

A final group expressed their predicament primarily in terms of concerns with achievement. This issue was a little different from underperformance, less a matter of efficiency or concentration or ability than of tangible success. Here the crucial question was something like, "Given that I am working hard, why can't I realize my plans and ambitions?" Some voiced disappointment that their lives were not working out the way they expected. Somewhat older on average, they do not have the financial resources, stability, or social status they feel they should have by this point in their life. Many were also unhappy with their relationships or that they were not married. A related group of participants, somewhat younger, felt harried by their circumstances. They were both financially strapped and drowning in things to do and disappointed that they were not attaining what they hoped and expected.[34]

For interview participants, addressing their predicament was a matter of answering such a question—a question about self and self-adequacy—and negotiating a path toward realizing the standard.

Analytically, we might think of the predicament as arising after the suffering itself, in the (subsequent) efforts to ameliorate or surmount it. But the experience of the suffering is itself part of the predicament, bound up with evaluations of self and one's relations to others, not something separate from the struggle to find meaning and make a response. What is wrong with me and what to do about it are tightly intertwined and mutually implicative issues, and any redescription can both reshape emotionality and reorder earlier judgments. What happens later is critical for how the primary phenomenon of the suffering is itself interpreted.

All aspects of a predicament involve interpretations. As I already noted, everyday suffering can only be described on the basis of experience. It involves self-understanding and status considerations, comparisons to others and evaluation in light of valued standards and ideals. These are not "raw facts" that can simply be read off experience in the same way glucose levels can be read off a blood test. Nor, as we saw in Kristin's story, are the medical concepts—diagnosis, disorder, medication effects, brain abnormalities—that participants used simply thing-like, independent of interpretive framework. These concepts too take their meaning from within the account of the predicament and efforts at its amelioration. Participant stories are inescapably acts of interpretation, a weaving together that expresses a current understanding of what experience means and how it is related to larger self-understandings and aspirations.

To inquire into predicaments, then, is to explore these acts of interpretation. This is itself a hermeneutical or interpretative task. In the interpretive tradition of social science and philosophy in which I am working, we as persons understand ourselves through our acts of interpretation, through the systems of meanings that allow us to make sense of our world.[35] Therefore, in studying the predicaments of others, I am not simply reporting their acts of interpretation, I am also interpreting them; I am interpreting acts of interpretation.[36]

## Interpreting Acts of Interpretation

In analyzing the interviews, I sought to enter into the participant's world. I sought to explore what people said about their experiences, about how their suffering had affected them, how it had originated, and what actions they had or were taking and why. I also sought to understand the emotions they identified and the ideals and standards against which they measured themselves and to which they aspired. I sought to understand their story.[37]

Considering what was *said* in the story and situating its meaning necessarily involved questions of morality and philosophical anthropology—questions of how we conceive of ourselves as persons in the world. What, I asked, makes particular interpretations compelling or moving, what gives these interpretations their force? As the anthropologist Renato Rosaldo has argued, "Cultural descriptions should seek out force."[38] That is, they should attend to the dynamism—persuasive, emotional, motivational—

informing explanatory constructions and patterns of action. They should attend in the matter of everyday suffering, I will emphasize, to the moral, to the relation of persons to the good. By this, I refer not so much to moral codes but to our sense of purpose and meaning, to our perception of what it means to be responsible, to our instinctive grasp of what is wrong and blameworthy.[39] In interpreting participant acts of interpretation, I sought to understand what normative ideals were in play, what system of right order was at stake, what type of self that it was good to be.

My method was comparative. A single interview has obvious limitations. Understanding is constrained by the relative dearth of personal background information, history, and context. The value is in breadth, allowing for comparisons across interviews and the identification of common patterns of meaning. While no two stories of suffering are the same—all suffering has an inherently personal and private aspect—the grammar and forms of explanation drawn on by participants are public. Their stories are framed in languages, metaphors, and models that others can understand and recognize. These patterns illuminate shared representations that are only visible when many individual stories are compared.

The patterns also point to background assumptions and ideals that both informed and shaped meaning for sufferers. Interpretive approaches presuppose a distinction between a "background" or "horizon" of meaning and explicitly affirmed knowledges.[40] In order to understand people's interpretations and their shared meaning-formation, this background, of which the participant might not be aware, has to be brought into view.[41] As the depth and breadth of the background is, by definition, limitless, no interpretation can ever be complete. There is always more context that could be brought to bear. I aimed to draw out at least that background that was necessary to make sense of the patterns I found in the stories. This background, as I will elaborate below, is both institutional and cultural.

Consider again Kristin's story. If not faulty biology, then we might be tempted to think that her problem is really one of faulty thinking. When psychologists write about the rising tide of struggles besetting college students, this is often one interpretation. When faced with the question, "If I'm not the best, who am I?" according to a book written by a former head of student mental health services at Harvard, some students mistakenly "decide, I'm a failure and a disgrace."[42] What these students need, psychologists like this tell us, are some "critical thinking skills" or cognitive behavioral therapy (in which problems are interpreted as responses to distorted thinking; why you think you're a failure is irrelevant) to correct their bad "habits of thought."[43] In this view, students—in mysteriously large numbers—have gotten confused and simply need to gain a more balanced perspective on their environment and on themselves. The problem is in their heads, and all that a student like Kristin needs is to adjust her unhelpful pattern of thought and aim for the right things.

But in an interpretive approach, talk of faulty thinking ignores or denies the background, something real and tangible in the world to which Kristin is responding. She is responding to real "norms of being" that are making real *moral* demands on her. Under our dominant regime of liberal selfhood, "potential" is a language of personal possibilities. That Kristin cannot meet her own standard, and suffers as a consequence, is inexplicable to her. It is, after all, *her* potential, and she insists, she does not "feel pressure from anybody to be someone that I'm not." What could possibly be stopping her from living up to it? She sees two options. On the one hand, she could try to "rescale" her potential downward to a lower status. But that would require her to change what it means for her to be a good person, since what is good and right for her is connected with her having *this* potential. On the other hand, perhaps something else is at work, something without the same status-lowering implications, something neurobiological. Those are the alternatives she perceives, and we can gain no purchase on the meaning of her predicament, her distress, or her response without taking the wider normative background into consideration.

Kristin herself believes that a problem in her biological makeup is the cause of her predicament. Part of the interpretative task will be to explore why participants like her think in such terms. Her belief, though, raises the question of neuroscience, about which, as I noted in the preface, so many claims of breakthroughs are now made. If brain malfunctions are the cause of problems like Kristin's, then perhaps talk of interpretation and the background are out of place. The diagnosis captures all that is important to know. Professional views of causation are now heavily biogenic but cover a range of explanatory models, from a "hard" medical model—"All psychiatric illness is best explained solely in terms of molecular neuroscience"—to more multicausal approaches that retain some place for social and psychological understandings.[44] In fact, professional views show many of the same polarities as lay understandings.[45] There is no one model and no definitive knowledge that can adjudicate. All behavior and mental activity have neurobiological correlates, and both nature and nurture condition us; it couldn't be otherwise. But there is no science that tells us that the causes of Kristin's "symptoms" are biological in nature.

And even if we could say a lot more about the brain at work, the question of interpretation would remain. Looking at a situation like Kristin's as though nothing but physical processes were relevant or at stake is to badly miss the mark. Predicaments, ethical quandaries, dark nights of the soul, and other challenges of the human condition do not reduce to physical matters. Of course, physical malfunctions might be present and significant (and surely are in the psychotic conditions). But before we can bring whatever we really know about such malfunctions into a meaningful conversation about ourselves and our suffering, then we will also need to systematically relate that knowledge with the other—and nonphysical—things we know about what it means to be a person. Otherwise, we end up, as in our current situation, with a lot of loose talk about the brain. This loose talk is having real and pernicious consequences, which is an important reason why the quiet shift toward the "neurobiological imaginary" needs to be explored—and challenged.

## A Shift in Imaginary

In interpreting the patterns of interpretation, my approach was also inductive. Behind the reasons participants gave, the emotions they identified, and the comparative standards they used, I found a yet more general pattern at work, a shift toward a new "imaginary." Of course, as an adjective, "imaginary," as in "imaginary friend," means pretend or make-believe. But the noun form, as used in social theory, refers to largely tacit and socially shared meanings that animate and legitimate common practices and are carried in modes of address, images, stories, symbols, habitus, and the like.[46] When I speak of an imaginary, I am not referring to professional discourse or theoretical understandings, nor to an abstract system of ideas or to some fully developed outlook. Rather, I am talking about an inchoate and not necessarily logical set of background assumptions that people draw upon to understand themselves and their place in the world.

Among ordinary people, I will argue, there has been a growing break with an earlier way of imagining suffering in terms of mental life and interpersonal experience, an imaginary we might broadly term "psychological" or "psychosocial." In its place are new intuitions, assumptions, and meanings about self, society, and suffering, and how they interrelate. This way of imagining I will refer to as "neurobiological." Its indicators, the subject of chapter 1, include imagining the genesis of suffering in mechanistic, typically biological terms, most evident in participant prioritizing of brain states—for example, "chemical imbalance," "misfiring brain circuitry"—as the effective cause of suffering. It is a way of imagining suffering without temporality, social context, or meaning, such as appears in Kristin's contrast of her brief experience talking with the counselor and her desire to solve her problem with medication. And it is a way of imagining predicament resolution that comports with a strong view of self as self-defining.

The neurobiological imaginary is the most general tendency that I found, the pattern behind the other patterns of meaning. While common, its presence is a matter of degree, and so I'll have to make distinctions between different groups of participants. As I work through the patterns in participant stories, I will relate them back to the imaginary and in so doing bring its appeal, practical and moral, into progressively clearer focus—as well as some of its many troubling consequences.

In the background of participant interpretations lies institutionalized medical languages and practices. Kristin, for instance, references common medical concepts, such as disorder categories, symptoms, and brain abnormalities. This is not her own language but a "grammar" she is drawing on to express, explain, and validate suffering and frame effective intervention. The source of this grammar is a biomedical landscape of meaning—a "healthscape" of ideas, symbols, and institutionalized practices—that has emerged since the 1970s. A confluence of specific and interacting developments were involved, both professional, such as changes to the DSM, and popular.[47] The upshot of these developments, the subject of chapter 2, has been a "biologization" of everyday suffering.

All the neuroscience talk in psychiatry has fed a widespread assumption that people look to medication and speak of their suffering in terms of neurobiology because they have been schooled in scientific knowledge. The interviews show this way of conceiving the relationship between professional and popular understanding to be deeply mistaken. While Kristin uses medical concepts, she did not get this language from her psychiatrist. She encountered it in other parts of the healthscape—media and everyday social interactions. Although she finds the ADHD diagnosis a persuasive explanation because it confers a reality on her experience and legitimates the use of medication, she also has her own view of what ADHD means, what a mental disorder is, what Adderall does and doesn't do, and so on. What the biomedical ideas mean and what makes them compelling to (some) people is not predetermined. They are not like a template that is simply laid over personal experience. In the interpretive struggle of a predicament, these ideas are appropriated under a meaning linked to the person's specific experience and aspirations. I will consider this "impression point" in chapter 3.

Disorder categories and taking medication, however, despite their practical and symbolic benefits, also carry negative connotations. Participants raised concerns about disclosure, diagnosis, medication, and neurobiological explanation, concerns often acknowledged in conventional discussions but attributed to issues of blame and stigma. But those were not the issues for participants. Rather, they were worried about implications of being "different," both in their own eyes and those of others. In order to understand and situate these worries, and the efforts participants made to contain or neutralize the danger, I have to bring into focus another part of the background, the broader cultural matrix of social norms and identity-values. Predicaments involve challenges to self, to one's being, and medical language and medication *add* to these challenges. Participant efforts to navigate them, to resist differentness, is the subject of chapter 4.

Participant resistance to medicalizing language and practice is not some negative struggle with stereotypes but a positive effort to restore their self-conception, the self that it is good to be. Exploring their evaluative standards and aspirations, the subject of chapter 5, illuminates the norms of "viable selfhood" that underlay and ordered—to various degrees, to be sure—how participants interpreted the nature of their predicament and the terms in which it might be overcome. In a complex society like ours, marked by many social differences and subcultures, there is a range of practices, authorities, and visions constituent of the good life. Yet, in the face of predicaments, I found common norms of being—efficient, optimizing, instrumental, and autonomous—that are widely valued in our society. With the goal of establishing or reestablishing this "viability," many participants, as we saw with Kristin, defined their predicaments as constraints on their volition, a constraint that arises outside them and apart from them. Recourse to the neurobiological grammar presented a way to imagine this constraint as an object that could be technologically addressed.

For participants who imagined the amelioration of their predicament in this way, the technical means is medication. In the relation between desired being and the objectivist means, the concern of chapter 6, there was a close affinity. Establishing or restoring viable selfhood did not require any qualitative elaboration of self, any incorporating of past history, any questioning of social circumstances. Steps that would have once been considered crucial can now, apparently, be bypassed. That such a course is even plausible signals a background change in sensibility, a reorientation to self, inner experience, and lifeworld. What we have now, following on social, economic, and technological changes, is a far more fluid social order that fosters a view of ourselves as "light" and psychologically transparent. It prizes a highly reflexive and present-oriented self-awareness. And it promotes an accommodating view of the social world as a (relatively) open space for self-designing. In this world of choice and autonomous selfhood, limitations seem irrational, arising from some alien place. The objectivist means "after psychology" address that alien presence and offer a way to establish or restore favored being.

The neurobiological, as revealed in participant stories, is a way to imagine ourselves, with the aid of techniques and technologies, as truly standing apart, to think of and enact ourselves or crucial aspects of ourselves independently of our situatedness, and to author our own story, free of constraints. Such is the promise and the appeal. In the final chapter, I'll consider some of the human costs, shifts in understanding that appear to follow when suffering is redescribed in mechanistic terms.

## Conclusion

What I heard the participants tell me, I believe, is that our dominant regime of the self is a primary source for much everyday suffering, that the kind of people we take ourselves to be makes heavy demands on us and carries unforgiving obligations. Our very image of freedom is creating a crisis of the spirit.

I will not offer an alternative regime or strategy for living. There are alternatives, and I will briefly highlight, in chapter 5, some of the participants who were living by other norms. Instead, my goal is to challenge a way of imagining everyday suffering that closes off alternatives, a way of imagining in thin, individualistic, and mechanistic—"neurobiological"—terms, terms whose effect is to naturalize the dominant regime of self and its grip on us. My argument is not with medication. As I will show, both historically and in the experience of interview participants, taking psychoactive medication does not require mechanistic thinking, nor does the possible value of medication for relieving distress depend on it. The interpretation of medication effects will be my concern, not medication or its potential value, as such.

Further, while I will be arguing for a picture of personhood that places a premium on dialogical relations and the inner life, and therefore on the possible value of psychotherapy or counseling, I will not be advocating for any particular form of such help. In fact, as commonly practiced, psychotherapists often foster the very regime of the self that I will be scrutinizing and can promote their own highly mechanistic view of suffering.[48] For a genuine alternative to the neurobiological imaginary, we need a more critical relation to the dominant norms of being and a deeper and more interpretive and qualitative view of ourselves.

# ONE

# The Neurobiological Imaginary

"The wiring, it's wackomo," says Lisa. That's the offhand way the thirty-year-old small-business owner describes her brain. But if it sounds like a joke, she is absolutely serious in her belief that "wiring" issues are the cause of the suffering she has dealt with for years. She didn't always think so. For a long time, and through much counseling, she thought her problems stemmed from "stuff that went on in my life." Now, with the steady encouragement and coaching of her primary care physician, she sees her emotional responses to stressful situations and a tendency to berate herself as indicative of a genetic condition—a slightly miswired brain. While the explanation seems right to her, it nonetheless chafes. She often finds it difficult to embrace.

As she tells it, her struggles began at the time of her parents' divorce, when she was in middle school. Close with her father, she took the separation very hard. To ease the pain, she began sneaking drinks after school ("self-medicating") and channeled her energy into her studies. She got good grades and was, she says, "very high functioning." In college, however, she started partying frequently and often drinking to the point of blacking out. During her junior year, she also began to realize she wasn't "very straight." Coming out of the closet, she lost her ROTC (Reserve Officers' Training Corps) scholarship, and her family more or less cut off contact with her. She began a downward spiral: insomnia, not wanting to get out of bed, crying for no reason. One night, in a cry for help to her friends, she took a few too many pills, though only enough, she notes, to "cause alarm."

Lisa's friends heard her cry and took her to the campus counseling center. She saw a counselor, who immediately referred her to the in-house psychiatrist for medication. The psychiatric visit was brief and formal: a discussion of symptoms followed by a diagnosis of depression and a prescription for an antidepressant. Lisa was initially reluctant to take the drug, worried that it would make her emotionally numb. The counselor reassured her, "You'll still be able to feel. It'll just be, it'll be what other people feel." Alarmed by those words, she wondered whether her emotions were somehow not what they should be, not genuine, too intense, and in need of change. Despite her reservations, she started the medication.

Today, despite the bad wiring, Lisa feels that she is finally getting her life together. She has reestablished contact with her family and has opened a small business. But the years between college and the present were bumpy, with many ups and downs. After graduation, she continued to take the antidepressant. She also continued to drink. Two years after college, she got drunk one night at work and things came to a head. She started attending Alcoholics Anonymous (AA) meetings, and getting sober made her realize that she had been using alcohol to deal with what she called "old stuff" from her past. She began seeing a counselor again.

Since college, Lisa has quit using the medication three times without telling her doctor or therapist. Each time she returned to old habits—wanting to stay in bed, self-criticism, picking fights—and her friends and therapist noticed and convinced her to restart the medication. This has confirmed for her that she is better on medication than off. Coming to that view has not been easy. She always felt that she should be able to manage her emotions and is ashamed that she seemingly cannot. Her self-described "contentious relationship with being on medication" has only improved as the understanding of her predicament has evolved.

Her continuing struggle with her old habits has opened Lisa to the idea that her struggles have a physical cause, the "biological stuff" that her primary care doctor kept telling her about. Rejecting the view that her old habits reflect internalized responses to external life events and unresolved inner conflicts, she now believes they are caused by malfunctioning brain circuitry, which is genetic in origin and triggered by difficult circumstances. Her struggles with her emotions do not mean that she is a failure or can't handle things; they simply indicate she has a physical illness like any other. Her doctor has helped her, in her words, "realize that if I had a heart problem I would take medication to make sure my heart is okay, and so I've got this nerve symptoms thing that's going on in the brain. It's not okay, so you're just taking stuff to make that better." The shift in "perspective"—a changed vantage point or vision—has led Lisa to explore her family history.

While she knows that she cannot diagnose her relatives, she thinks that some people in her family exhibit symptoms of depression, reinforcing her belief in the biological source of her problems.

Though still ambivalent, Lisa strongly credits the medication for helping her to function better, be more rational and controlled, and conduct herself more ethically. She has had some bad experience with side effects, and she also believes that her concern with losing emotional range has been somewhat confirmed. She wonders if her sadness and her happiness are now both stunted and whether losing emotional intensity has undermined her motivation and ability to face aspects of her life. But medication, she says, helps make her emotions more efficient and creates a time lag between stimulus and response: "It's helped me be less reactive so that when something bad or something good happens, I'm able to just sort of sit with that for a second before reacting, because in the past if somebody said something a little confrontational to me, I would right away, I'd get really upset." On the medication, by contrast, she can hear hard things from others and realize "this is about them; this isn't about me." And she can see that her own self-criticism is irrational.

Concerned that she is overstating the role of medication, Lisa stresses, more than once, that "it's not all the meds." She is no longer seeing a therapist or going to AA. But she values the counseling she has received and the social support of AA in getting sober. She believes they have helped her in many ways to take more control over her circumstances, be more assertive in her relationships, and increase her self-esteem. Her antidepressants have helped a lot, but, she clarifies, "they're not like a magic drug."

---

In Lisa's story, her changed perspective on the cause of her suffering is central. About her experience in the wake of the divorce, coming out, and other difficult life events, she actually says little. Instead, she concentrates on her history of treatments, beginning with the drinks as a child and leading up to her seeming inability to get free of medication. What is inexplicable to her is that she still cannot, unaided, regulate her emotions in the socially expected way, and this reflects on her as a person. Her predicament—the question she needs answered—concerns this lack of control, and she has come to see the explanation in terms of a misfiring brain.

Among all of those interviewed, Lisa reported one of the most challenging efforts to negotiate a predicament. She has had a hard time reconciling the norm and ideals against which she evaluates her experience—authenticity, efficient emotion, and self-sufficiency—with each other and with the use of medication. For Lisa, the idea that the drug brings her emotional state into the normal range raises the question of what makes her unmedicated emotions less real or less illuminating of her experience. She asks, "How is my experience not true?" Does the loss of emotional richness, even feeling bad, not constitute a loss of something important? Is not a medicated mood somehow fake? Here, Lisa judges herself in terms of authenticity. While this ideal means different things, as I'll discuss later in the book, for Lisa, it seems to center on a getting in touch with her feelings as a guide to her truth, rather than merely conforming with social conventions. She may have acquired this orientation in therapy or AA. The other participant with a similar orientation had also been in both psychotherapy and AA for several years. She, too, struggled with the authenticity implications of taking medication.

Against her authenticity, understood in this way, Lisa has confronted conventions of appropriate emotional expression—"what other people feel." Throughout the interview, whenever she referred to changes in her feelings, whether worse when off medication or better when on, she also stressed that other people, both her therapists and her friends, notice. While she certainly likes being less reactive and conducting herself better, she also gets the message that there are particular feeling rules to which she is subject. Under these rules, she must operate in a restrained and efficient emotional register, neither more nor less intense than average and keyed to external social expectations and not necessarily anything inside. This is just the register that she credits the medication with helping to produce. It gives her a disengaged perspective. She can distance herself and view interactions almost as though they were happening to someone else. On medication, she can follow the rules better. But for Lisa, complying with feeling rules runs counter to authenticity as a standard of rightness.

While authenticity means that the emotions expressed are *her* emotions, it does not necessarily imply any lack of emotional control. For Lisa, self-mastery is critical, and she evaluates her continued use of medication against a standard of self-sufficiency or autonomy. By this standard, she should be in control of her emotions by now. Her need for medication represents a kind of dependence that calls her competence into question. This is why she experiences shame. Her continued drug use seems to show her up as the type of person who cannot get her life together. This is the conclusion, the inexorable logic, she has struggled hard to escape.

Lisa has come around to the misfiring brain explanation because it quiets all of her worries at once. It affirms that she is not being inauthentic, for the strong and unruly emotions she experiences are not actually hers. The emotional register of those with a "normal" brain is the appropriate standard. In fact, it is the *only* standard. And self-control is not even at issue; it cannot touch what is in fact wrong. She is dealing with a physical disorder, and there is no shame in continued treatment. It is what her viability as a norm-conforming person, one who is autonomous and properly authoring her own life, requires.

Lisa's story illustrates, in a pronounced way, a change of interpretation that was common among the participants. The change is from what I will call a "psychologizing" to a "medicalizing" form of explanation, from, roughly, an account of suffering in terms of the mind to an account in terms of the brain.[1] What makes Lisa's story stand out is that she has a very positive view of her experience of counseling and AA (which she calls a form of therapy) and resisted a change of perspective for so long because she had a strongly psychological outlook. Compared to the other people who changed their interpretation, this was somewhat unusual. They were less psychologically oriented to begin with, and this was another reason why Lisa was one of those who had the hardest time with the change. Her transformation, however, is helpful for bringing into relief the tension between the two explanatory tendencies.

In this chapter, I present initial evidence from the interviews for the new imaginary I am calling "neurobiological." Like Lisa, many people described shifting to the view that their suffering was primarily (though not necessarily exclusively) the result of a neurobiological aberration. Their effort to ameliorate their predicament, which might have begun with some counseling or psychotherapy, centered on medication. One indicator of the imaginary was this change to a medicalizing vision. Another was in the more general ways in which participants spoke that often also pointed toward a mechanistic, nonpersonal way of imagining the cause of suffering and the appropriate course of action to take. Other studies of public and patient views toward questions of problem cause, medication, and psychotherapy have shown similar medicalizing views, and I conclude by briefly considering these.

## Psychologizing and Medicalizing Perspectives

It is important to emphasize again that I am talking about the views of ordinary people, what are often referred to as "lay" or "folk" (i.e., nonprofessional) understandings. By design, none of the people we interviewed was a practicing psychologist, psychiatrist, or other medical or mental health professional (though two were in graduate school; one in social work and one in psychology). Their stories reflect their own evaluations and understandings, based on their experience, sources of information, and background conceptions.[2] Participants identified various sources of influence on their thinking, to be sure, including, for some like Lisa, professionals whom they had seen. But their interpretations, for reasons I will show, should not be confused with professional discourses or read as efforts to reproduce them.[3]

What I am calling the "psychologizing perspective" is admittedly a catchall category, covering orientations or visions that were only in a few instances given a name, whether "psychological," or "new age," or "religious." Into this category I put anyone who did not explicitly reference the brain or heredity as the cause of problems. Generally, these people spoke of some aspect of themselves—problematic tendencies in the mind or personality or in emotional reactions to others—that led to their predicament by making the circumstances they were facing very difficult to handle. Only a few expressed a distinctly psychological rationale. These few interpreted their current troubles as arising from formative events in their past, such as abuse and family issues. Because these events were stressful, however, their emotional import was hidden from awareness or consciously avoided. They remained unresolved and continued to have an impact on emotional well-being. The key to improvement for these people was to gain an understanding of the role of these earlier events. The majority, by contrast, spoke in more indeterminate terms and did not reference any specific psychological theory or mechanisms (e.g., intrapsychic conflict, unconscious motivation). They were "psychologizing" in only a very general sense. They regarded their suffering as involving some (undefined) nonrational factors and did not see it as the direct result of intentional action.

For those who spoke from a medicalizing perspective or vision, as we saw in Lisa's changed outlook, the primary locus of trouble was in the brain, and reference was made to somatic factors—a "chemical imbalance" or "nerve symptoms thing," to use Lisa's phrase—that influence thought, mood, or behavior mechanistically. They regarded this faulty chemistry or wiring, and the vulnerability to suffering that it creates, as very likely inherited, though the problem might have gone unrecognized for a long time or laid dormant during periods of favorable circumstances. Once triggered, the brain malfunction was outside their volitional control and did not involve any conscious beliefs or desires. The mental states (thoughts, desires, emotions) it produced, they implied, were without content or object. They were arbitrary, unpredictable, and not "about" anything.

In Lisa's story, the medicalizing perspective has slowly come to replace the psychologizing, at least with respect to the predicament that she recounts in the interview. For her, the choice is either one or the other. And as typically described by those who changed, they are antithetical visions. However, in many instances, people who emphasized a biological cause of their suffering retained some meaningfulness by including other factors they saw as relevant. Ella's story is a good example. At the time of the interview, she was working various temporary jobs and hoping to return to school. Now thirty, she was first diagnosed with depression as a teenager, shortly after her mother's death. Despite some resistance she has felt in the African American community, which she describes as unsympathetic to the reality of mental illness, she has come to the view that her suffering is caused by "a chemical thing in the brain." But she has attributed some of her feelings to the event itself, the tragic nature of her mother's death, and to her natural tendency to take things hard. Ella, like some others, takes a medicalizing stance but qualifies it with some reference to psychologizing or circumstantial factors.

Incidentally, Ella's comment about resistance in the African American community introduces one of the few ways I found in which demographic factors affect interpretive practice. I didn't mention it at the outset of the chapter, but Lisa suggests that another reason she has had trouble coming to terms with the faulty wiring explanation is her working-class upbringing. Her family, she says, stigmatized help-seeking and taking medication. Starting in college, she was surrounded by people from the upper middle class, and her attitude changed. Some others also spoke of different social class orientations. The clearest difference, however, was among the African American participants (sixteen people), regardless of social class. Many stressed that there was a stigma in the African American community toward matters of mental health. They suggested that this stigma could lead to a denial of everyday suffering and made recourse to either psychotherapy or psychiatrists more challenging (though only one indicated that it actually stopped him from taking a desired action).

Ella's story shows how a medicalizing perspective, though centered on brain states, could be enlarged with additional factors. The reverse might in theory also be true. A psychologizing vision might make room for some ideas about the mechanisms of the brain. But I saw little evidence for such a standpoint in the interviews. In only two cases did someone with a psychologizing outlook also make mention of a problem with their brain. In each case, the person had begun medication, and the vague brain references appeared to signal an initial movement toward a change of perspective. The strong indication from the interviews was that once a person had begun to speak in terms of the brain, other ways of thinking were marginalized if not eclipsed. Even the small number of those with a medicalizing perspective who had more or less given up on any positive help from a medication did not then change their perspective. Changes of view only went in one direction.

### Orientations to Ameliorative Action

We might think, given the contrasting emphasis in the two perspectives toward mind or brain, that they would, in turn, be directly correlated with corresponding efforts at amelioration. The picture was more complex. There was a range of possible responses to predicaments. People employed different vocabularies and practices in different ways and in different combinations. Some possible combinations of explanation and intervention seemed to entail logical incompatibilities. No one, for instance, explained their predicament in the medicalizing language of brain aberrations while also engaged in intensive insight-oriented psychotherapy. But for the participants, the links between explanatory vision and action were not matters of abstract logic. The associations they made emerged in the effort to make sense of personal experience. These could be quite idiosyncratic, and it was clear that how accounts of suffering origin and preferred practices might be related was not inevitable or predetermined.

Only the few mentioned earlier, with a distinct psychological orientation, addressed their predicament in classical psychotherapeutic terms. In their psychologizing stories, they had worked through a connection between their life history and their emotional struggles, faced their real, underlying issue, and begun to cope with its effects in their lives. They described this introspective task as challenging because it meant taking a hard look at their life, facing up to fears, difficult emotions, and unpleasant memories, and refusing to "put up walls" in self-protection. Therapy helped them take control, change habits of thought, and identify triggering events that cause them trouble. Interestingly, two of these four people had also been on medication, which they found provided some needed, temporary relief. But it was no substitute for self-insight. Sooner or later, as one, Hassan, put it, "you have to work through it on your own." In a sample of eighty people, these four were something of an outlier. But at least at one time in American history, they would have been a significant presence, and their diminishing numbers, confirmed by other evidence that I will cite below, is a sign of the shift I am exploring.

Paradoxically, the people with a psychologizing perspective, which was about half the sample, were the *least* likely to have ever pursued any sort of psychotherapy. As I am using a very general sense of "psychologizing"—any account in terms of life events or experience of self that did not explicitly reference the body/brain—so by "psychotherapy" I am including counseling or talk therapy of any type or with any sort of professional (psychologist, social worker, physician, clergy, etc.). When we asked these participants if they had ever sought any sort of help, we prompted them to include any counselor or therapist they had ever talked to, even if only once.[4] About one-quarter had done so, for whatever length of time, and most of these were ambivalent about the experience and did not seem to have much interest in or hope that introspection and talk would improve their situation. In fact, half were taking a medication, which they credited with providing some relief and taking the edge off of bad feelings. Some of those who had never seen a therapist or counselor expressed a modest interest in seeing someone, while others were indifferent and even opposing. What really stood out, across the board, was a distinct lack of enthusiasm and even a somewhat greater openness to medication.

Most commonly, the people with a psychologizing vision were coping on their own. In their stories, many stressed their own efforts to deal with their suffering. Some believed that improving their situation was a matter of accepting their limitations or avoiding troublesome contexts or engaging in efforts at self-care. They did not blame themselves for causing their suffering, though some saw change as a matter of greater personal effort and right attitude, even if they did not necessarily know what to do or how to reorient or improve themselves. For example, Phil, age fifty, became quite withdrawn and isolated after an accident put him in a wheelchair for eight months. He has not been able to reestablish his relationships and sees his unhappiness and isolation as partly his own fault, "just the lack of doing something about it." Like others, he does not say that his inaction or poor coping is intentional, much less that it reflects something like weak character or lack of self-control. What it does reflect, at a minimum, is the belief that his own efforts *should* be adequate and that he *should not* require the help of others. For Phil, there is apparently nothing to discover about his experience. He, who is averse to trying medication, is "not a big sharing person," and regards self-help techniques as "stupid," has become somewhat fatalistic about his situation.

Whereas the psychologizing addressed their predicaments in a variety of ways, the people with a medicalizing perspective were almost all taking a medication. And this was the group that included most of those with experience of psychotherapy or counseling. Unlike Lisa, however, they did not incorporate much of that experience into their stories. Rather, they focused on their medication experience, about which they were generally positive. They credited the drug with at least some general improvements, such as less moodiness or anxiousness and increased confidence, energy, or concentration. Where differences emerged was in their attitude to the drug. Some, particularly those who initiated the request for medication, stressed the crucial difference it was making in their lives. Others emphasizing their ambivalence toward taking it.

The ambivalent, like Ella, attributed their suffering to a biological cause but qualified this explanation with some reference to psychologizing or situational factors. They also reported a more conflicted relationship to the medication, worrying about having to rely on a chemical, losing control, and experiencing undesirable effects. As a result, they moderated their use of the drug in various ways—taking it less often than prescribed or stopping and restarting it—or expressed a desire to get off the drug in the near future. Being on a medication was a threat to their sense of independence.

A conflicted relationship to medication, like Lisa's struggle with a change of outlook, illustrates some of the particular challenges raised by a medicalizing form of explanation. It required certain kinds of "boundary work" to negotiate meaning and contain undesired implications. In her story, Lisa negotiates the boundary between what in her emotions is mechanistically determined and what is *hers*; between what can be controlled and what is outside control; between what medication changes and what remains untouched. For Lisa it was especially challenging, but everyone who attributed their suffering to brain states engaged in boundary work around the issue of mechanistic causation. There were additional challenges as well, including negotiating the medical languages of diagnosis and illness/disorder and the problem of disclosure to others. All raised implications of "differentness." I will discuss these issues in detail in coming chapters.[5] Those with a medicalizing perspective had a lot to say about them. So too did those with a psychologizing vision, who often referenced such challenges as reasons for rejecting medicalizing interpretations.

I will have far less to say about the specific challenges of a psychologizing explanation because participants did not speak directly of them. When most people spoke of or about counseling or psychotherapy, for example, they didn't regard it with much seriousness or expect much from it. Recall Kristin from the introduction, the student who began taking Adderall without a prescription and who was struggling to live up to her potential. In talking about her three sessions with a counselor, she mentioned that there were things that she was "afraid of inside." The facing of these things suggests a demanding feature of counseling. Kristin herself is dismissive, giving as her reason for quitting that she was only "going to yammer on about certain things," which would be "too much of a hassle" and unlikely to achieve anything "very lofty." The few people who spoke in classic psychotherapeutic terms certainly noted demanding features, describing introspection and disclosure to another person as difficult and requiring from them candor, an openness to confront painful emotions, and effort to put emotional experience into words.

What can be said of the challenges of a psychologizing explanation will emerge over the coming chapters indirectly, as those features that a biological explanation can bypass or avoid. Lisa's account in terms of neurobiology, for instance, allows emotional experience to be detached from her *self*—made nonreflective of her and her evaluative stance—in a way that her previous psychologizing account did not allow. The psychologizing language suggested something unresolved and, in effect, set limits to disengaging experience, to imagining herself as free of her predicament.

## Toward a Medicalizing Perspective and Medication

As this brief review shows, no single explanatory format or appropriate intervention commanded common assent. Still, there were patterns, and they were not random. They suggest certain assumptions, certain ways of imagining, that are held in common. I begin with the many people who changed their explanation. All shifted toward a medicalizing perspective, and I will consider this shift in relation to their interaction with clinicians, nonmedical therapists, and medications. These interactions reinforce the perspective, but they do not determine it.

### Changes of Perspective

In about half the total interviews, participants explained their predicament in medicalizing terms. Like Lisa, many reported a change in perspective over time, and this change was always toward biological causation. Prior to the change, they framed everyday problems in terms of mental life or personality or circumstances, which was presumably the initial default for everyone with a medicalizing perspective whether they reported a change of view or not. They changed perspective in the face of disorientation, at the point of suffering and a predicament, when their self was called into question.

Most people who came to take a medicalizing perspective had done (or were doing) three things: they had seen a doctor or other clinician, attended sessions with a psychotherapist or other nonmedical counselor, and taken a medication. Not all—there were a few with this vision who had done *none* of these things, separately or together. And there were some who spoke in psychologizing terms who *had*. Neither a doctor visit nor any specific type of treatment is always accompanied by a change of mind. But they did often go together, raising the question about the role of doctors, therapists, and taking medication in promoting understandings of problems in terms of malfunctioning brains.

The majority of those who gave a medicalizing explanation had seen a psychiatrist, general physician, or nurse practitioner and been diagnosed with a mental disorder. Many also reported that their clinician spoke of their diagnosed condition as arising from an aberration in their neurochemistry or genetic makeup. People understood these doctors to be promoting medicalized understandings of their suffering. In the next chapter, I will explore the biological turn in psychiatry and the emergence of new theories, languages, and practices for describing and acting on everyday suffering. And in chapter 3, I will consider how and where ordinary people come in contact with such language. In that context, I will say more about what people reported about physician promotion of medicalized interpretations and how they received them.

It is enough to note here that a person like Lisa, who describes an ongoing conversation with her doctor about biochemical causation, is the exception. Most of those who drew on biological language—and not all of these had even seen a doctor—reported learning about this explanation elsewhere. According to participants, the main reason doctors spoke of chemical imbalances or faulty wiring was rhetorical. Clinicians made analogies with physical problems when encouraging the patient to begin a medication or when trying to overcome their reluctance to stay on it. Their point was that the patient had a (minor) physical problem and should take medication for it just as they would for other physical problems. While some participants found this talk helpful, most did not portray their doctor or nurse practitioner as playing a defining role in how they interpreted the cause of their suffering or as engaging with them in any wider discussion of self and meaning. The medical interview did not require such roles, and unless the person expressed some resistance to the proposed medication, the diagnosis and prescription proceeded without any need for a concurrence of shared meaning.

### Exposure to Psychotherapists

We might expect brain talk from psychiatrists and other doctors. Nonmedical psychologists, social workers, and other counselors are another matter. From them we would anticipate an emphasis on psychologizing explanations, not medicalizing. How to make sense, then, of the fact that of the people who explained their suffering in terms in neurobiology, three-quarters had some experience with individual therapy/counseling (plus one in group therapy), either at an earlier time or currently at the time of the interview? Of those with a psychologizing vision, by contrast, only about one-quarter had ever attended any individual therapy (none had attended group therapy). Exposure to therapists was nearly as highly associated with a medicalizing explanation as exposure to medical doctors.

I did not find evidence that therapists and counselors directly promote medicalizing perspectives. The one exception perhaps proves the rule. During her junior year of high school, Becca's physician, who had diagnosed her with depression and prescribed an antidepressant, recommended she also see a professional counselor. She went, she says, to a couple of sessions, but reports that the counselor told her that she didn't have any psychological issues to work through other than being "type A" and "a bit hard on herself." According to Becca, her counselor felt that her real problem was that her brain chemicals were, in her words, "screwed up," and that talking about her depression would do little to help. Needless to say, the counseling ended.

Psychotherapeutic change depends on exploring and reconsidering the meaning of personal experience. Therapists guide change by using verbal techniques, symbols from their therapeutic rationale—psychodynamic, cognitive behavioral, and so on—and emotionally arousing talk to lead clients to decode, reinterpret, and reexperience their specific problems in light of the new meanings.[6] There is no point to such therapy if persuasion like this cannot be of any help. But Becca's report is quite unusual. None of the others who had seen a therapist or counselor said anything like this. While most spoke sparingly about what actually transpired in their therapy sessions, what they did say—getting advice and support, airing feelings, leaning behavioral "strategies," and so on—reflected a psychologizing not a medicalizing orientation.

If therapists do not promote medicalizing understandings, many, such as Lisa's, do encourage drug-taking.[7] Most of the people with a medicalizing perspective had seen *both* a counselor and a medical practitioner.[8] An important reason why is that in half the cases a therapist recommended also seeing a physician or a physician recommended also seeing a therapist. In all these cases, participants explained their suffering primarily in terms of brain malfunction and took a medication.[9] And most did not stay in or continue with their therapy for long. By contrast, there was only one person who reported being sent by her therapist to a medical professional (where she received a prescription for a medication) and who retained an account of her suffering in psychologizing terms.

In the final analysis, by far the best predictor of a psychologizing vision was *not* to have seen any type of professional. Of those who had never seen a doctor or a therapist, virtually all explained their predicament in terms of the mind. Since a person's explanatory account can be formulated before or independent of professional contact, we cannot read too much into these correlations, by themselves, as influencing changes of perspective. As we saw with Lisa, change requires motivation, and professional influence has to be set within the larger nexus of the predicament and the types of amelioration that seem possible and desirable in light of questions of self-worth. But it does seem safe to say that psychotherapists and counselors are a weak influence on preserving or promoting a psychologizing outlook. At least one reason is suggested by another significant correlation in the interpretive patterns.

### Exposure to Psychotropic Medications

Of those with a medicalizing orientation, the clear majority were also taking medication at the time of the interview. This association raises the possibility that taking a drug might lead one to reinterpret one's suffering as arising from a physical malfunction in the brain that the drug puts right.

There was evidence for such a feedback loop. While the judgment that a biological aberration was at the root of problems could come prior to or apart from taking a medication, it was more prominent after. In some cases, participants cited the positive impact of the medication as support for a medicalizing account. Hailey, for instance, twenty-seven years old, first saw a psychiatrist when her anxiety about speaking in groups and talking on the telephone was interfering with her productivity at work. She doesn't remember the initial visit very well but recalls answering a lot of questions and the psychiatrist recommending an antidepressant. The drug, she felt, "genuinely worked," and so, she recounts, "I think that lent support to my thinking it was a biological thing." And later in the interview, Hailey adds more strongly about her anxiety, "it's definitely something I needed medication for. So, it's something biological."[10]

This reasoning was sometimes apparent even when the participant felt the drug was not working very well. And it was apparent even in the absence of drugs. Some concluded that if the remedy is *not* physical, then the problem must not be physical either. Luis, for example, twenty-two and working, has had serious difficulties with attention since he was a child, but says he will only be ready to believe he has "an actual physical issue" if his attempts at rising above his predicament—reading self-help books and advice from friends and family—fail. Since they are currently helping and he does not believe he requires medication, he is not prepared to say his problem is neurobiological.

Not everyone who took medication adopted a medicalizing vision (almost one-third did not). Most reported a fairly positive evaluation of drug effectiveness (itself a complex issue to which I will return in chapter 6), but taking medication did not change how they understood the genesis of their suffering. Gladys, for instance, fifty-two, was unemployed at the time of the interview after the manufacturing business she had been building for five years had to shut down for financial reasons. She's been diagnosed with depression and takes two antidepressants. While the drugs help some with sleep and low energy, Gladys sees her struggles in psychologizing terms and emphasizes the help she gets from positive thinking. She reads "lots of self-help books," and adds, "That's what I guess you would say our religious philosophy is based upon: that you have the power to change everything and do everything and live a healthy, happy, fruitful life." For her, taking medication is quite compatible with her philosophy.

Like Gladys, people acknowledged some emotional and behavioral improvement with the medication but did not then draw the conclusion from this improvement of a physical cause. It has long been claimed that the success of drugs in the treatment of mental disorders constitutes a powerful reason to believe that disorders have a biochemical origin. This was what the psychiatrist Peter Kramer famously concluded from his exercise in "listening to Prozac."[11] The very use of a drug—a chemical substance—speaks to and convinces people of neurochemical causation. It is what psychologist Lauren Slater in another best seller, *Prozac Diary*, meant when she wrote of Prozac having a "belief system" that "only chemicals can cause hurt, and thus, only chemicals can cure."[12] The witness of participants, however, suggests that drug and belief system are not joined as she presupposes. Drugs, in themselves, do not necessarily or automatically lead to reductionist inferences about physical causes. Further, as I will discuss in greater detail below and in chapter 2, a whole generation of people before the 1980s took prescription psychotropic medications with little or no reference to neurobiological explanations. Then, as now, the relation between drug and causal explanation is not one of necessity or inevitability.[13]

To those already predisposed to attribute their suffering to brain problems, there is little question that the practice of taking a medication can open up or confirm them in that perspective. This is not a matter of feeling some imperative to logically square a physical cause with a physical remedy. Rather, with the brain explanation, the medication can now appear as evidence of and a response to the brain problem. Lisa, with whom I began this chapter, was on medication from the start of her story and only later adopted a medicalizing perspective. What changed her mind was not the medication but the biomechanistic explanation by which she could detach continuing emotional volatility and self-criticism from herself. By a feedback loop with this new understanding, the medication then became desirable in a new way and a critical part of what made the talk of faulty wiring appear plausible. It could change what she seemingly could not.

This feedback loop was not present in every case. Some who took a medicalizing perspective were not even on a medication. But it was an important element in most stories, with implications for drug evaluation. But with these remarks, I am getting ahead of the argument. In coming chapters, I will have to draw on further evidence to show that the feedback loop moves in the direction I am suggesting.

## Other Evidence of a Medicalizing Orientation

That half of the participants have adopted a medicalizing perspective, and, further, that most of them are on medication, tells only part of the story of the trend toward the neurobiological imaginary. There are other indicators that the lay environment is increasingly saturated with medicalizing language and expectations for dealing with everyday suffering. One was in what people said about psychotherapy, both those with therapy experience and those without. For the most part, participant talk of psychotherapy was marked by meager descriptions of benefit or low expectations of help. This talk was in contrast to views of medication. Even most of those with a psychologizing vision appeared to see "real" troubles as matters for a doctor, not a therapist, and for medication treatment, not talk.

### The Low Expectations and Influence of Psychotherapy

Half of all participants had been to some individual therapy or counseling, ranging in duration from two or three sessions to one year or longer. Their experiences varied. In some of the stories, people described their therapy in somewhat negative terms: they didn't find it very helpful or get much out of it. In all but one of these cases, the participant was currently taking a drug. In a similar number of stories, participants describe attending a few sessions of therapy and deciding that was enough. They generally report getting some good advice or benefiting from the opportunity to freely share their feelings. In most of these cases, the person stopped once they began taking medication. On the drug, they didn't seem to see much need for or purpose in further talk.[14]

In the balance of the cases, which involved somewhat longer therapy, participants generally portrayed therapists as a kind of helpful friend or even coach, but seldom as any kind of interpretive authority. Only a few people, for example, said anything about their therapist's interpretive framework or rationale, the conceptual system the therapist uses to explore and encourage change in the meaning of client experience.[15] One participant, Donna, age twenty-five, for instance, who had been seeing a cognitive behavioral therapist for about nine months at the time of the interview, referred to the cognitive behavioral therapy (CBT) strategies as "tools to handle life." But she does not share the CBT view that her depression is the result of distorted thinking.[16] Earlier, she had seen a psychodynamic therapist. Psychodynamic therapy locates problem origin in intrapsychic processes or conflicts. Donna does not see her problems in those terms either and complains that her "so Freudian" psychiatrist was not giving her useful feedback but "taking notes for four years." While psychotherapy is a technique-driven process, it has often been noted that its success depends in part on the client adopting the therapist's theory of causation for the deficits that the person is experiencing.[17] There was not much direct evidence of such adoption in these stories, or even awareness of the operative theory. While at least seven people had been in some form of CBT, for example, five took a medicalizing view of their problems.

Just a few people, like Lisa, credited their therapy with help in ameliorating suffering. Generally, they recounted their therapy experiences in positive yet fairly anodyne terms. Many emphasized that it was helpful to have someone with whom they could share. Daniel, for example, age forty-eight, wrestles with being antisocial and has problems with authority. He went to a psychiatrist for counseling at his sister's urging after he had an altercation with a police officer following a routine traffic stop. He explains, "We had numerous sessions. I saw him [the psychiatrist] on a weekly basis for I guess it was six months." Although he "really can't recall the conversations," he does report that after the sessions, "I feel a little better about myself."

Some of the same sense of low expectations comes across in the remarks of people who had never attended any therapy. If they did not volunteer comments, we asked, "Have you ever considered going to see someone?" In their responses, often brief and sometimes dismissive, many indicated that they had never given therapy or counseling serious consideration or any consideration at all. The out-of-pocket cost of therapy, often noted in public discussions as an obstacle to psychotherapeutic help-seeking, was not the issue.[18] Nor did it appear that people found the prospect intimidating. The research literature of a generation ago described going to a therapist as a significant threshold. In his 1969 book *Why People Go to Psychiatrists*, Charles Kadushin, for instance, writes "going to a psychotherapist, even if no treatment at all takes place, changes a person's self-concept from that of a person who is by and large self-sufficient to that of a person who requires professional help to get through life."[19] Acknowledging the need for such help was challenging, and the path to help-seeking involved overcoming considerable resistance. But not here. Those interviewed, for the most part, did not see much stigma in going to psychotherapy. They seemed to share the view of Gretchen, aged thirty-five, who said, "Therapy is now like not an embarrassing thing."

There were a few exceptions, mostly men.[20] The other people without therapy experience expressed no such reservations or concerns. Seeking help from a psychotherapist was simply not important. This indifference was in sharp contrast to remarks about drugs. Those who did not take medication almost always explained why, even with no prompting. In general, people seemed to regard medication-taking as the obvious, expected intervention that, in one way or another, demands a response, whether positive or negative. Not so with psychotherapy.

### Presumptions of Medication Effectiveness

Medication enjoyed a far better reputation than psychotherapy. Many of those taking a medication believed it was helping them, and many of those not taking a medication were open to or actively considering the possibility. For example, Amber had been wrestling with stress, humiliation, and a sense of failure after her fiancé, and the father of her two children, walked out on her a couple of months before the interview. She is currently dealing with the predicament on her own, but if she does not feel better within "a couple months or in another month even," she is ready to think "there's something probably wrong with my head," and to seek medical help. She clearly anticipates that the help will be pharmaceutical in nature, saying she would begin by talking to her friend Julie, because "she's just like a walking pharmacy."

Even many of the people who expressed some fairly strong aversion to medication (nine men, seven women) indicated that they would probably take it if their situation warranted it or a doctor so advised them. They too believed that the drugs were likely to be efficacious in ameliorating problems. Their objections came for other reasons, such as fear of it changing their personality or identity or having adverse side effects. Maya, for example, a college sophomore, says she has been "sad," "unhappy," and "chronically not up to par" since starting high school. She attributes her feelings to social isolation and academic pressure and acutely feels she is not living up to her potential. Maya considers her feelings of sadness to be a serious problem; she thinks she is probably "clinically depressed." But, she says, she cannot admit that she is depressed because it would mean she is incapable of solving the problem herself and will need to take drugs. Maya doesn't want to use drugs. As she bluntly puts it, "I don't want to be clinically depressed because that means that I'll need drugs or something and that is horrible," or worse "cosmically screwed." But her objection to drugs concerns dependence; she assumes they can help her "out of it."

Similarly, some people equated "real" emotional and behavioral problems with the need for medication treatment. One example is Andrew, a college sophomore, who reasoned from the fact that he was prescribed medication to the conclusion that he has "real problems." Andrew struggled with profound feelings of emptiness after his father abandoned the family when he was in high school. He was ultimately sent by his mother to see a psychiatrist. At the first meeting, he was diagnosed with depression and prescribed an antidepressant. He said:

> It was kind of life altering when he told me that I would need to be prescribed medication. It was more like an affirmation that you actually have real problems. They are not stupid and petty problems like you keep saying they are. They are not things that you can simply just talk out. . . . When he told me I needed medication I was kind of like, oh, okay, so I do have real problems. It's not just everyday stuff that all people deal with.

The opposition that Andrew sets up nicely captures a common contrast people made between the low expectations of psychotherapy and perceptions of medication efficacy. There are things you can "just talk out," or "everyday stuff that all people deal with," and for these things therapy might be of some value if you are so inclined. Then there are the "real problems"—with medical names—and dealing with these is a question of medication.

Pointing toward this very dynamic, the British sociologist Nick Crossley observes, "Whatever 'new age' maps of the soul we are attached to in our moments of reflection, we are increasingly drawn to bio-medicine in our moments of need."[21] The moment of need is the moment of disorientation and suffering, the moment when the "everyday stuff" takes on the shape of a real problem.

## The Neurobiological Imaginary

From their accounts of the genesis of their suffering and their response to it, I found that participants, to varying degrees, conveyed three distinct tendencies in the face of a predicament. The first was a tendency to shift from an initial or default explanation of suffering in vernacular terms of mental life to an account in terms of malfunctioning brain states that mechanistically disturb thought, emotion, or behavior. The second was a tendency toward an anemic or indifferent attitude toward insight psychotherapy or counseling and a meager investment in any form of such help. If a professional was needed to ameliorate or rise above a predicament, it was a doctor. The third tendency, whether taking a drug or not, was to regard medication as the necessary and effective intervention to ease symptoms and address "real" problems. While these tendencies are not a package and do not logically imply one another, many people expressed all three, a substantial majority at least two, and some one. Only a very few expressed none of these dispositions.

This constellation of tendencies reflects the way of imagining that I am calling neurobiological. Direct explanation in terms of heredity and the brain is the clearest indicator. But views of counseling and the special efficacy of medication were also revealing for some of those who did not reference brain abnormalities. These people too seemed to have little confidence or trust in a first-person, qualitative language of mental life or imagine their own struggles in terms of any inner experience or normative conflict. Many conveyed the impression that in their experience there was nothing to explore, or discover, or confront, and some sought to separate their emotional experience from themselves. Like the medicalizing, many contrasted a soft, subjective world of talk and therapy with a hard, objective world of medicine and medication. They pitted "real" problems against problems that are "in your head," where "real" always seemed to mean somatic and medication the logical intervention.[22] Some avoided help-seeking because they worried that they might have or a doctor might think they have a "real" problem in this sense. Some "psychologizing" participants, in other words, imagined suffering and its proper response in essentially medicalizing terms even when they did not invoke a biological causal language about their current experience.

Moreover, there are additional indicators of the neurobiological that I have not yet had a chance to discuss and will need the following chapters to draw out. These include the end point toward which people aimed when seeking to ameliorate or rise above their predicament—what I will call a "viable self," a self that is workable and passes normative muster—as well as the very language in which they told their stories. One of the most striking features of the interviews was the typically thin, impersonal vocabulary people used to talk about themselves and their experience. The English language has more than four hundred emotion words. It was remarkable how few of these words the people we interviewed actually used. Though this was no doubt partly due to the interview context, their evaluative language often had a distinctly mechanical, behaviorist quality to it, with many references to being functional, efficient, rational, productive, in control, and so on. They typically expressed their predicaments in a vocabulary of difference and deficiency—inadequate, insufficient, imbalanced, not measuring up. Across the interviews, participants spoke of life as a project of one's own devising and directing. A quotidian life, yet one in which *everything* matters and for which compliance with dominant norms of being is both a compelling good and the sign of a good self. For this self-making, reference to one's history or inner life was marginally relevant at most. The implicit picture of the good self did not require it.

I emphasized above that the tendencies in explanation and action were with respect to the specific challenges of a predicament. While predicaments are a critical window on meaning-making precisely because suffering and deep commitments—distinctions of worth by which we define ourselves—are at stake, they are only one context of experience. I speak of "tendencies" because they are part of the flux and change of an ongoing negotiation of the meaning of experience. Participants drew upon different types of explanation and action orientation, some coexisting in the same story, that reflected different experiences, different needs, different hopes, different ways of containing undesirable implications for self-understanding. There is a discernible trend in the interviews but no single way in which people understand themselves.

I use the term "imaginary" to emphasize that while I am talking in part about ideas that people expressed, it is not "ideas" in the sense of a conceptually articulated knowing or an explicit belief system that were revealed in the interviews. Rather, in the "perspectives" they conveyed, the words they used, and the actions they took or didn't take, participants communicated thoughts and impressions and assumptions about how society and suffering, the self and the body (brain) might fit together. An imaginary is a way of imagining our situation, how it is going and how it *should* go, a socially shared way of seeing that enables sense-making about the world and an orientation to our place in it.[23] In an hour of need, depicting ourselves in a mechanistic language of the brain or absent an inner complexity or in need of a medication to feel and perform like other people has taken on a new plausibility.

But is the constellation of tendencies I found merely the sum of the idiosyncratic dispositions and attitudes of a nonrepresentative group of volunteer interview participants? To answer this question, I sought out nationally representative surveys of patients and the general public. They show a similar medicalizing configuration of views. Studies that inquire into how people who receive mental health care explain their condition and their treatment show that explanatory factors such as heredity and chemical imbalances have grown in relative importance over time, as has an emphasis on medication.[24] A similar trend also appears in broad public surveys of views about mental disorders and help-seeking. Especially relevant are "time-trend" analyses, which compare responses to identical items at two or more times with a sample drawn from the same population. Time-trend studies in the United States document a shift among members of the general public toward regarding mental health problems as rooted in or influenced by biological factors and toward an increased endorsement of medical treatment and prescription medicine.[25]

Findings from this research are broadly consistent with the patterns that I found in the stories of interview participants. There has been a discernable turning away from older conceptualizations of mental health problems in terms of psychological and social causes and psychotherapy treatments toward biology and medication. Study authors charting this trend expect it to continue.[26]

To be sure, all of these studies continue to show significant endorsements of psychosocial factors and of psychotherapy. In fact, other studies that have surveyed people about their "preferences for psychological versus pharmacologic treatments for psychiatric disorders" have often found a higher preference for psychological treatment.[27] This preference (which is lower for people who have actually sought treatment) seems somewhat at odds with the trend I have described. Some, noting such preference studies, have blamed insurance reimbursement practices, uninformed doctors, and the pharmaceutical advertising juggernaut for keeping people from getting the psychosocial treatments they actually want.

There is no question that as a percentage of overall treatment, psychotherapy has played a rapidly declining role in mental health care as "a

large and increasing proportion of mental health outpatients received psychotropic medication without psychotherapy."[28] During the period from
1987 to 2007, when the number of Americans who were being treated
for a mental health problem rose dramatically, the percentage of Americans who use psychotherapy on an annual basis remained flat, at just over
3 percent.[29] Further, those who do receive psychotherapy receive less and
less of it. In the decade up to 2007, the average annual number of therapy visits declined by nearly 20 percent. The great majority of clients are
now effectively receiving a therapy of very limited duration or focus.[30] The
more open-ended, introspective psychotherapy of yesteryear, not to mention old-fashioned psychoanalysis is disappearing.[31] Brief or short-term
therapy—once viewed as "a second-rate, 'Band-Aid' treatment to be used
only in cases of expediency"[32]—is now the norm.

Of course, as some argue, there are important institutional factors at
work contributing to these changes.[33] But the fact is, as clinical psychologists regularly note, client demand has also changed. While it has probably
always been the case that most people who seek psychotherapy are not interested in long-term treatment, "consumers' wishes for quick results" have
notably intensified.[34] Apart from changes in behavioral health care and
professional practice, clients themselves are also driving the trend toward
very short-term treatments.[35] This too is a sign of the growing break with
the past.

## Conclusion

What stands out from the interpretive patterns of how everyday suffering
is understood and addressed is the growing ascendency of explanations
and expectations tied to biology and medication at the expense of alternatives. The shift toward a neurobiological imaginary is not just observable
in the stories of those who have adopted a medicalizing perspective but
in what the other participants indicate about understandings of problems
and their resolution. Further, a displacement of the psychologizing is not
some spurious finding of my nonrepresentative sample. Time-series studies
show the same distinct trend. And they show how much has changed in
recent decades.

To understand this shift and what has propelled it, we have to consider
background understandings in both a deeper and more immediate range.
The neurobiological imaginary is not an intellectual scheme or theory; it
constitutes a horizon of meaning within which ordinary people imagine
their predicament and its resolution. In the deeper background are cultural

changes in norms and ideals, in ways of understanding ourselves. While I will allude to this level throughout, I will take it up more directly beginning in chapter 4. In the more immediate background is a historical change in the "healthscape" of contemporary mental health philosophy that I will call the "biologization" of everyday suffering. This change, which has arisen in the interplay between available drugs, psychiatric theorizing, commercial forces, and popular representations, is the subject of the next chapter.

# TWO

# The Biologization of Everyday Suffering

Eric, thirty-three, was on the verge of a meltdown. He was a new teacher at a rigorous elementary school in a wealthy suburb and was struggling to meet deadlines. He often felt unprepared, under pressure from the principal, the intense parents, and "a maniacal co-teacher," and blamed for things that were not his fault. These dealings were especially stressful because, he says, he is not well organized, is thin skinned, and interpersonal relations are not his strong suit. As the year wore on, he felt himself growing more anxious and short-tempered, with increasingly disjointed thoughts and trouble sleeping. He was "going nuts" and wanted to quit.

Seeing his distress, Eric's wife recommended he see a physician they knew. Given the crisis at work, Eric readily agreed. Already knowledgeable from other (unspecified) sources, he also had a pretty good idea of what to expect. He thought the medical visit would be straightforward because he thought his problem was obvious and would be obvious to the doctor. And, in fact, he reports, after a physical, they had a casual conversation, and she quickly made the anticipated diagnosis of generalized anxiety disorder (GAD). Next, she wrote him a prescription for an antidepressant, as well as a sedative, as needed, for sleep. Except for some worry about nausea, he believed the drug would be good for him, and he had no hesitations about starting it.

According to Eric, the doctor presented the diagnosis and medication in a very reassuring light. She drew analogies to common ailments and pointed out that, while nonmedical approaches may be helpful, they may be insufficient. Specifically, he said,

> She was really helpful and sort of making the idea of it not seem . . . like, you're crazy, you need meds. Making it seem more like, if you had a head-

ache, you'd take Tylenol. Some people have blood pressure problems. Some people have X, Y, and Z. Based on what you've told me and your history, you've got issues with anxiety. And sure, it's going to help to exercise, it's going to help to go to yoga, it's going to help to do all this stuff, but it may not be enough, especially if you're not self-medicating. So, she was nice in twisting the idea in a positive light.

Moreover, she told him that the medication would help alleviate his anxiety by "keeping the serotonin in your brain available to you."

Soon after, Eric decided that he would quit his job. He began feeling better on the antidepressant, describing its primary effect as giving him some emotional detachment and perspective. It enabled him to step outside of the stress at school and made finishing the year less difficult. At the time of the interview, Eric had been away from teaching for two years and was a stay-at-home father. He is still taking the antidepressant and thinks he may be on it for the rest of his life.

The prospect of long-term treatment seems appropriate because Eric now thinks, in light of the diagnosis and discussion with the doctor, that an anxiety problem is something he has had all his life. He just didn't recognize his childhood experiences of nervousness and fidgeting—even vomiting in certain competitive situations—as a problem because he was young and it somehow seemed normal to him. Similarly, he didn't understand the frustrations of his work life prior to teaching as reflecting a problem with anxiety.

Eric was a high achiever in college and, after graduating, enthusiastically threw himself into a series of jobs. Each time he was disappointed. In his twenties, for instance, he cooked professionally for several years and set his sights on being a "top-notch chef." He had a lot of drive and got very good. But, he says, he didn't always make the best choices, and the work was frustrating and exhausting. He eventually left this career disenchanted and intent on quitting the rat race. It was then that he decided to pursue teaching, expecting something more laid-back. That he struggled with the teaching too is a sign to him of a chronic anxiety problem. And he interprets his history of regular use of nicotine and marijuana as instinctive acts of self-medicating to manage it.

Previously, he thought of his anxiety as simply a part of his personality or disposition, and as reflecting, as he puts it, his "finely tuned nervous system that doesn't quite settle." Now, in light of explanations from his physician and his own reading, he holds a broadly neurochemical account of his anxiety. He also adds in further support that there is a family history—

three of his sisters are on an antidepressant—and he offers an analogy from "yogic theory" to illustrate how his anxiety might ultimately be of somatic origin, "maybe even genetic, passed down." For Eric, a lifelong condition may require a lifelong treatment.

The possibility of indefinite treatment does not disturb him for another reason. Despite speaking of "twisting the idea," he agrees with the doctor's assurance of the ordinariness of his treatment and diagnosed condition. The antidepressant, which he compares to an asthma inhaler, is "just another medication." Before the trouble as a teacher, he knew little about anxiety disorders in the same disinterested way he knew little "about oceanography." Now he knows that the GAD diagnosis is a recognized medical condition, and it helps him to see that his problems are not "like me just being whacked out in my head or nervous or stressed about stuff that's ridiculous." Still, he doesn't actually accept the idea that he has a mental disorder—that would be something far more serious, like schizophrenia—and distances himself from the formal category of GAD. That is what the doctor calls it. His own word is simply "anxiety."

Two years on, Eric is doing a lot better. He gives some credit for his improvement to situational changes, especially the end of teaching, and to his increased yoga, exercise, and gardening. But the medication has been crucial. While he can "function comfortably," not all of his anxiety has disappeared. The doctor has periodically increased his dosage, and on at least one occasion, she resisted his desire to go even higher:

> For a while, I was thinking, what else? Maybe I need to up my dosage even more. And in the end, we were talking and she's like, you know, it's not a magic bullet. And I was like, you're right. If it was a magic bullet, everyone would be on it. Every single person in the world if it was a magic bullet that made everything awesome. It just helps me cope.

Yet even if his anxiety is still "a little more than a normal person," he remarks, he is amazed "that a white pill, how can this do this?"

Above all, he credits the medication with slowing down his emotional responses and allowing him to disengage himself from his previous aspirations and sense of obligation. He specifically mentions living up to his parents' expectations, worry about letting people down, and feeling "those sort of testosterone pressures. You know, like the macho man?" He now sees these not as reasons for his anxiety but symptoms of it, symptoms that the drug relieves.

For this reason, Eric believes that on the medication he is not his old self. Rather, he is a "better person," who also communicates his feelings better and has better habits, like smoking less marijuana. He now believes that people in his position *should* take medication, if, he says, "they [are] aware of the situation and have the willpower. . . . If they could bring themselves to do it and identify it as a problem external from whatever's going on in their life."

---

As we saw in chapter 1, there is clear evidence for a growing break with a time when both the lay public and those receiving treatment saw the roots of behavior and emotional problems in psychosocial (psychologizing, interpersonal, environmental) reasons and causes. New medicalizing ways of describing and acting on personal troubles have increasingly come to the fore. When people speak of diagnostic categories like GAD and attention deficit disorder (ADD), or when they locate the roots of their suffering in a chemical imbalance, or when they express a fear that acknowledging their struggles to a professional will lead to a prescription for medication, they are attesting to a new "healthscape" of deeply institutionalized medical concepts and practices, not of their making and often endowed with a powerful facticity. This healthscape is a context, enabling and constraining, with which people in predicaments must contend. Alternative languages, explanatory frameworks, and practices remain available, to be sure. Even those who take a medicalizing perspective sometimes mix it with other factors. Still, the new context influences the pathways of care and the parameters of what people might recognize and credit.

The account that Eric gives of his experience suggests many of the new elements, as well as the kind of narrative that can now be told about the relation of persons to their selves. He had his own language and understanding of his predicament. He speaks not in terms of illness but in a vernacular: he was "revved up," had "a meltdown," has a "thin skin," and is "sort of flighty and nervous" in the face of stress. His way of talking links his anxiety to his personality, not to a condition he suffers from. The doctor, by contrast, according to Eric, speaks of symptoms, disorder, medications, neurochemistry. It was not a vocabulary that he previously knew, and he hears it as an obscure language of science, like "oceanography." However, he came prepared to try medical management and was "game" to try the medication. He was suffering and seeking to escape his predicament.

Eric explicitly credits his physician with interpretative assistance. In one sense, this meant translating his "anxiety issues" into the language of disorder, a language that competes with and at least partially displaces his vernacular talk and ideas. In the redescription, his experience is given a more medical and treatment-receptive meaning. Medication is indicated, while other efforts at amelioration are either ruled out (nothing is said about psychotherapy, and Eric has never considered it) or ruled insufficient (yoga, exercise, "self-medication"). His anxiety is also "real" in a new way: it is no longer peculiar to his personality or his circumstances, but an independent phenomenon that affects many people. As he says, "this is recognized." It has an ontology "external from whatever's going on in" life.

At the same time, in contrast to colloquial "anxiety," talk of the recognized disorder of GAD is talk of mental illness. A second type of interpretative assistance that Eric attributes to the doctor aims to mitigate this association. She helped normalize his experience by comparing it with the common sort of physical problems doctors treat and by emphasizing the ordinariness of psychotropic medication. As he tells it, her comments about serotonin were in this context of putting a "positive light" on his situation. Rather than indicate or reinforce any notion of "being whacked out in my head," in Eric's words, reference to "the neurochemistry" is part of a positive prognosis; it identifies the fixable problem for which the drug is the necessary technology.

Perhaps most clearly and importantly for Eric, the new medical language offers him a new narrative of self, one he did not have available previously. With the doctor's assistance, he reduces his past experience to a single general feature. His seemingly disparate teaching-related stress, history of disappointments, and marijuana use are emplotted as symptoms of the same underlying thing. This telling establishes not only a more medicalized picture, potentially genetic and chronic, but it opens up an avenue for aspects of self—the things he used to care about; norms he desired to meet—to be disengaged from himself. In his new account, those aspects of his emotional experience and evaluative outlook that he now disowns are the manifestations of biological anomalies. On the medication, he is free of them, more relaxed and sensitive to others—he is a "better person," committed to asking, "How are you going to better yourself?"

The terms of Eric's story reflect the neurobiological imaginary and the picture of the good that it opens up. We cannot understand this way of imagining self, the reasons for suffering, and the means to be delivered from it without some history of critical changes that have taken place over the past half century.

In this chapter, I explore the shift in mental health philosophy and practice regime through a series of historical twists and turns. My account begins with the serendipitously discovered psychotropic properties of nonpsychotropic drugs and ends with the claim that the future of psychiatry is in neuroscience. While psychiatric research is one of the important actors in this story, much of what was theorized and hoped for has proved elusive or mistaken. Nonetheless, a decisive turn was taken, philosophical and practical. Everyday suffering, which was first psychologized, has now been "biologized," and both the practice of psychotherapy and the use of medication have changed profoundly. While the turn I describe involves the movement toward a biomedical orientation in psychiatry, my concern here (and continued in the next chapter) is not just with changes in psychiatric philosophy and practice but with the larger healthscape—the field of ideas, symbols, assumptions, actors, and institutions—that exists at the intersection and commingling of psychiatric and popular domains.[1] Both domains were at work simultaneously, and the influence on lay imaginary is in their mutually constitutive and contingent interaction. This healthscape provides the grammar of the neurobiological imaginary, shaping new social expectations, accessible interventions, and the languages available to express and validate suffering.

## Freud and Drugs for the Mind

While biological theories and somatic treatments date to the beginnings of psychiatry as a medical specialty, in the years after World War II, psychoanalytic ideas took center stage in the United States.[2] The years from the war until the mid-1960s constituted a "'golden age' of popularization" of Freudian thought, and the origins of psychological problems were sought in childhood experiences and efforts to adapt to environmental stress.[3] Experience during the war was important to this ascendency and its synthesis with the more environmentally oriented and psychosocial views of the leading American psychiatrist, Adolf Meyer. Both psychiatric experience with stress-related problems and public perception of psychiatric problems were very high during the war.[4] Further, many European psychoanalysts fled to the US, influencing practice here, while Nazism, not to mention problems with the existing somatic treatments, did much to discredit biological explanations of character and behavior.

However, it is important to note that while psychoanalytic and psychosocial orientations dominated psychiatric theory—shaping, for instance, the first (1952) and second (1968) editions of the *Diagnostic and Statistical Manual of Mental Disorders*—and were popular in the academy and media, treatment was more eclectic and pragmatic. The number of trained psychoanalysts during this period was in fact never large. In 1957, during the heyday of Freudian psychoanalysis, there were less than 950 analysts in the US, including some 700 members of the American Psychoanalytic Association, 140 members of the William Alanson White Institute, and about 100 members of the American Institute for Psychoanalysis.[5] The total number of psychiatrists and clinical psychologists was also small. In its final report, *Action for Mental Health*, the Joint Commission on Mental Illness and Health, first funded by an act of Congress in 1955, decried the serious "professional manpower" shortage facing the country. According to the report, there was only one psychiatrist for every fifteen thousand Americans in 1960. Most of the eighteen thousand psychologists in the country were college teachers or research employees of the federal government—"only about one-third," the Joint Commission wrote, "engage in clinical services, where their skills would count in the care of the mentally ill." The commission saw this as a small number in a country of some 180 million people.[6]

Further, somatic interventions continued to dominate treatment for the most serious disorders, such as those found among institutionalized patients. The first "tranquilizers," reserpine and chlorpromazine (marketed in the US under the brand name Thorazine), came to market in 1952, and by 1960, drugs such as these had moved to the forefront of treatment for psychotic patients. At least initially, the drugs were viewed and marketed not as replacements for but as adjuncts to psychoanalysis or psychotherapy.[7] In roughly the same period as these drugs were introduced, the population of mental hospitals, which had grown steadily, began to shrink.[8] One study suggested that the incidence of long-term psychodynamic therapy might have reached the peak of its popularity by about 1957.[9] In the following years, chemical treatments would become the norm for serious mental illness, with or without talk therapy.

Somewhat ironically, given its often dim view of somatic therapies, psychodynamic psychiatry played a significant role in laying the groundwork for the widespread use of psychotropic drugs in the general population.[10] For Freud, as sociologist Philip Rieff observed, "the commonplace is saturated with the abnormal, the pathological"—"psychopathology no longer deals with the exception but with the ordinary man."[11] One of Freud's key ideas is the dictum that "we are all somewhat hysterical," meaning, in Rieff's words, "that the difference between so-called normality and neurosis is only a matter of degree."[12] Accordingly, in his theoretical framework, Freud distinguished between two broad classes of disorders. The "psychoses" are the serious mental disorders, like schizophrenia and manic-depressive (bipolar) illness. These are very debilitating afflictions, typically involving significant distortions or misinterpretations of reality and personality disorganization, and only a relatively small fraction of the population suffers from them.[13] The "neuroses," by contrast, are the worries, psychological problems, and psychosomatic complaints of otherwise normal people. Freud believed these problems, rooted in anxiety, to be very common among the general public.

Several mid-century developments seemed to confirm his judgment: not only the experience of soldiers during World War II and the influence of Meyer, as noted earlier, but also a rapid increase in admission to psychiatric hospitals and outpatient clinics and the findings of large-scale epidemiological studies after the war. The Midtown Manhattan Study, for instance, conducted in the late 1940s and early 1950s, found a very high rate of stress symptoms, with over 80 percent of survey respondents reporting some mental health problems, mostly related to anxiety.[14] This study, and others like it,[15] emphasized a psychosocial perspective: anxiety was a psychic response to particular social settings and demands. The investigators of the Midtown study regarded many of the problems they found not as forms of mental illness but as expected and even adaptive responses, given the typically stressful work and family conditions in which people lived. Even so, the nonspecific distress and accompanying somatic symptoms were painful, and the Midtown researchers saw a large potential demand for professional intervention to help.[16]

Psychoanalytic theory provided the rationale and the impetus for this intervention. It established the pathological significance of everyday neurotic symptoms and emotional distress and made them appropriate targets for intervention. It also established a necessity to do so, since, according to the theory, these problems could lead to serious mental illness if left unchecked. Just such a perception of significant unmet and ramifying need animated the *Action for Mental Health* call for a national effort to dramatically boost the mental health workforce. The US, according to the final report, was in the midst of a mental health crisis that had to be addressed—a call that did not go unheeded. After 1960, the number and types of psychotherapists grew exponentially, with most of the growth occurring outside of the medical profession, including clinical and counseling psychologists, social workers, marriage and family counselors, and a large range of other sorts of guides and support groups. The era of "psychological man," as Philip Rieff predicted in the late 1950s,[17] was inaugurated, with a vast increase in professional help-seeking for "anxiety reactions" and other problems of generalized stress. In a time-series study, first conducted in 1957 and then in 1976, for instance, researchers concluded that "it is clear from the data that men and women have become much more psychological in their thinking about themselves and attempting to understand their own lives." The most important sign of this new way of thinking was the "very large change" in the use of mental health professionals, reflecting a "national investment in such expert help as a resource for conquering obstacles to the good life."[18]

### The Psychopharmaceutical Revolution

The vast expansion of the purview and practice of psychotherapies was one consequence of the ascendency of psychodynamic and psychosocial theory and the pathologizing of everyday distress and "obstacles to the good life." Another consequence, wholly unexpected, was to pave the way for the medical and popular embrace of new classes of psychotropic drugs that began to come on the market in 1955.[19] When the major tranquilizers, like chlorpromazine, arrived in the early 1950s, they proved effective in treating patients with psychoses and were later classed as "antipsychotics." While initially also marketed for all sorts of nonpsychotic problems—"severe emotional upset of the menopausal patient," "the child with a behavior disorder," "agitation," "the emotional stress that complicates so many somatic disorders," and much more[20]—they had serious side effects and were mostly associated with the treatment of debilitating, chronic mental illness. They never gained much cultural visibility. Hard on their heels, however, came another class of sedative compounds, less powerful than the antipsychotics. These drugs were reportedly safer than the barbiturates, which had been the most popular everyday sedatives in prior decades.[21] The barbiturates, however, had fallen out of favor because of the belated discovery that they could lead to physical dependence.[22] To avoid the bad name associated with "sedative" and to capitalize on chlorpromazine's favorable press, the new drugs were dubbed, for marketing purposes, the "minor tranquilizers." The first of these medications was meprobamate (brand names Miltown and Equanil), a compound whose muscle-relaxing properties had been observed while searching for something completely different—new antibacterial agents.[23] Massively advertised for anxiety problems in the general population, Miltown became the first psychiatric blockbuster and was, in its time, the fasting-selling prescription drug in American history. People flooded doctors' offices seeking prescriptions, and in the early years, shortages were common as production struggled to keep up with demand. Several dozen "me-too drugs" (i.e., drugs with nearly identical chemical structures) followed, and within two years of the release of Miltown, 7 percent of Americans indicated that they had ever tried "pills called tranquilizers."[24]

While psychoanalytic theory established the pathological significance of emotional and mental distress, its tool for addressing the root causes, psychoanalysis, involved insight and talk. Neither theory nor practice permitted much role for a pill. The goal was not to immediately *relieve* symptoms but to get at their *meaning*. But in promoting Miltown, its marketers, according to the historian David Herzberg, "echoed Freudians in emphasizing the ubiquity of anxiety even among otherwise healthy people."[25] They stressed a prophylactic therapeutics offered by regular physicians, which extended psychodynamic goals to a far larger population. General practitioners, after all, were the ones presented with most of the psychosomatic complaints and problems of nervousness and anxiety among ordinary people, not psychiatrists or other mental health professionals. These physicians could not psychoanalyze patients, but with an "antianxiety" pill, they could offer temporary relief and perhaps prevent a minor problem from becoming something worse.

Further, marketers cast Miltown not as a challenge to psychotherapy but as an adjunct, one that facilitated, in the words of a Miltown ad, "psychotherapeutic rapport."[26] The minor tranquilizer marketers were not alone. The stimulant medication methylphenidate (brand name Ritalin), which came on the US market in 1956, was originally advertised for the treatment of chronic fatigue, mild depression, narcolepsy, and other problems (its use with hyperactive, and then inattentive, children came later). As with Miltown, ads in the late 1950s promoted it as an aid to psychotherapy: Ritalin could "break down" a patient's "resistance," make him "more cooperative," and render "him more accessible to psychotherapy by promoting verbalization of repressed and subconscious material."[27] This way of putting therapy and drugs together was widely adopted, according to Herzberg, "within both psychiatry and medicine at large. Conventional wisdom, repeated in countless textbooks and instructional articles in medical journals, held that psychiatric drugs did not cure but allowed 'only a degree of symptomatic relief.'"[28] Yet this relief could be a critical step toward effective talk therapy by allowing the patient to "actually face the origins of her problems and work through them."[29] Drugs thereafter were often used in combination with psychotherapy. Within a shared psychosocial framework, using a medication and talking to a therapist could be complementary.

With the opening provided by Freudian ideas and their medical authority, advertisers for the minor tranquilizers, the stimulants, and the new antidepressants (imipramine and iproniazid were introduced in 1957) broadened the types of distress that merited medical treatment. Print ads in medical journals promised relief for a very wide range of daily problems, from stress, insomnia, "tired mother syndrome," "breakups," "marital tensions," and "the blahs" to all manner of aches and pains. Much of the marketing of these prescription-only drugs targeted doctors, from journal ads to the free samples and elegant dinners provided by pharmaceutical sales representatives. But the drug companies also worked to educate and influence prospective patients as well. While the Food and Drug Administration (FDA) prohibited direct-to-consumer advertising, the historian Andrea Tone writes, "nothing proscribed planting stories about the new drugs to magazines and newspapers, a strategy that helped achieve the same results."[30] Ad agencies directed public relations campaigns aimed to circulate in the popular media and generate buzz. They regularly provided information about the drugs to journalists and offered educational interviews with company representatives, fed tidbits about celebrity endorsements to gossip columnists, and orchestrated various publicity stunts and gimmicks. Glowing articles about the new wonder drugs were prominently featured in mass circulation news (*Time, Newsweek*) and women's (*Cosmopolitan, Ladies' Home Journal*) magazines.

In some cases, these stories were ghostwritten by industry. Herzberg cites two examples published in *Cosmopolitan* in 1955 and 1956. The first praised Miltown for "relieving stomach distress, skin problems, 'the blues' and depression, oversensitivity to summer heat, fatigue, inability to concentrate, lack of 'social ease,' behavior problems in children, and insomnia." The follow-up article provides another list of uses, including banishing "that tired feeling," making workers "happier and more productive in their jobs," and helping "frigid women" respond to their husbands.[31] As with ads in medical journals, the public relations campaigns touted the effectiveness of the drugs for distress symptoms and troubles with social roles and even promoted use to improve performance and gain a competitive edge. And, like the ads, stories often discussed everyday troubles without any reference to illness.

By 1960, three-quarters of physicians were writing prescriptions for one or more of the minor tranquilizers,[32] and in a survey that year, 14 percent of Americans indicated that they "ever [i.e., at any time in their life] had occasion to take a tranquilizer."[33] Those numbers would swell in subsequent years, following on the phenomenal success of a new group of minor tranquilizers. Amidst concerns over its side effects and addiction potential, Miltown began to lose its luster and was replaced on the blockbuster list by a class of compounds known as benzodiazepines. The first was chlordiazepoxide (brand name Librium), which came on the market in 1960; and the second was diazepam (brand name Valium), introduced in 1963 (many others would follow). In a 1971 study, 15 percent of Americans (20 percent of women and 8 percent of men) reported using a minor tranquilizer in the past year,[34] and by 1972, virtually 100 percent of family physicians and internists were prescribing them.[35] Valium would become, according to historian Tone, "the most widely prescribed pill in the Western world from the late 1960s to the early 1980s."[36] In the peak year of 1973, physicians in the US wrote 104 million prescriptions for the minor tranquilizers,[37] which were fast acting (thirty to sixty minutes) and typically used for only short periods or intermittently as needed.[38] Other prescription psychotropic medications, including the stimulants, antidepressants, and hypnotics, while not as popular, were also prescribed. In the 1971 study, a total of 8 percent of Americans had used one of these drugs in the preceding year, bringing the use of all prescription psychotropics to 22 percent of the adult population.[39]

For the most part, the pharmaceutical companies did not run afoul of psychodynamic practice when they presented drug indications in terms of general problems and distress arising from various life situations and crises. Psychodynamic theory places great store in early life experiences, unconscious conflicts, and maladaptive defense mechanisms that give rise to symptom patterns, and psychoanalysis and the insight psychotherapies it inspired are directed to exploring and resolving those psychic conflicts individual by individual. Neither theory nor practice specified or required clear, reliable categories of disease or disorder.[40] Broad talk of "minor emotional disturbances,"[41] or, to quote the 1979 *Physicians' Desk Reference* entry for Valium, "tension and anxiety states resulting from stressful circumstances or whenever somatic complaints are concomitants of emotional factors," could flourish.[42]

The actual prescribing patterns of physicians followed this lack of diagnostic emphasis and specificity. In the first study of its kind, conducted in 1959, researchers examined the prescriptions written by family physicians and specialists who were part of a New York group health plan. They found that "about one-third of the minor tranquilizers were for psychiatric disorders, the rest being distributed over a wide range of conditions."[43] Most of these conditions were somatic complaints that were likely to have accompanying psychic distress or to have been adjudged at least partly psychosomatic.[44] Some 12 percent of the prescriptions did not state any diagnosis at all.[45] A decade later, the situation was similar. According to a study of physician indications for Valium prescriptions, "Only 30 percent is in identified mental disorders and the remainder is in conditions that could be labelled as 'psychosomatic.'"[46] Even as late as 1989, a diagnosis of a mental disorder was only made in about one-third of the primary care visits in which a Valium-like drug was prescribed.[47]

Although critical pushback was beginning to grow, there was a widespread consensus that the popular drugs were calm-me-down or pick-me-up pills and could work for anyone. What they offered for the many and protean challenges of modern life was symptom reduction or improvement, and this was valued for its own sake and (on occasion) as an aid to *psychological* intervention. In one of the first studies of patient experiences with Valium in the 1970s, respondents identified conflict over their ability to perform a role or adapt to its demands as the primary reason why they continued taking the drug. Their reasons for starting medication, which were different, centered on social stress, "internal tensions," and somatic complaints with related social stress. None were reported to have spoken of a diagnosis, a mental disease, a chemical imbalance, or other such biomedical or biological talk.[48] By then, in historian Tone's words, people had long regarded it as "okay to see doctors for drugs to make them feel better about the vagaries of life, not just to treat diseases."[49]

What we have in the postwar period, then, is the emergence of new ways of thinking about and acting on behavioral and emotional problems arising from experiences of everyday life. The preeminence of Freudian theory in America and the appearance and popularity of new psychiatric drugs combined to greatly expand the scope of personal problems regarded as appropriate for doctors to treat. From the 1950s to the 1970s, both the use of psychotherapy and the use of prescription psychoactive drugs soared. Much of the talk and the prescribing were for common stresses and struggles and underwritten by a broadly psychosocial perspective with little reference to specific disease states or somatic explanations. And as far as the available data indicate, the fact that the drugs were prescribed by doctors did not lead people to the conclusion that they were suffering from a physical illness (consistent with a view of problems as situational, the large majority of patients who used the drugs did so on a short-term and episodic basis). Nor did the fact that the drugs were chemical agents lead them to the conclusion that their struggles were chemical in origin. The way that drugs were advertised and publicized reflected the same presuppositions.

But if general practitioners and patients were not "listening" to psychotropic drugs and hearing neurochemical causation, other professionals were. These professionals, who would come to represent a "neo-somatic movement,"[50] included not only biologically minded psychiatrists but biochemists, neuroendocrinologists, geneticists, and others. They saw the effectiveness of the new drugs as challenging existing conceptualizations of psychopathology and representing an opportunity for psychiatry to increase its status in medicine and make real treatment progress. Over the tranquilizer era, their "organic school of thought" would come into ever sharper conflict with the dominant psychoanalytic school and, in the 1970s, begin to substantially displace it.[51] The language of disorder and the strategy of treatment would fundamentally change.

## Drug Effects, the DSM-III, and the New Biological Psychiatry

The use of psychotropic compounds for mental health problems launched a new science, and, in time, a new psychiatry. The discovery in the 1950s that some compounds had effects on psychiatric symptoms had come about largely indirectly, by observing the "side effects" of drugs being used or tested for other purposes. Chlorpromazine, for example, the early antipsychotic, is an antihistamine that was tested for a variety of medical purposes, including as a preanesthetic to reduce potential shock from surgery. In the course of the clinical trials, researchers noticed that it reduced excitement and agitation, which led them to test it on schizophrenia patients. In another case, the observation that some patients on iproniazid, an effective tuberculosis medication, experienced increased vitality and euphoria led to testing and then use of the drug as an antidepressant. Other drugs followed a similar pattern. They were not developed based on any prior scientific knowledge of the etiology or pathophysiology of specific psychiatric conditions. No such knowledge was available. Consequently, while researchers documented beneficial effects from certain existing compounds, they possessed no direct evidence of how the drugs might actually achieve those effects.

For the biologically minded psychiatrists, the reported efficacy of the drugs strongly suggested that specific chemical processes might be involved in giving rise to mental illness. It was "the astonishing success of physical methods of treatment of these conditions," observed Alec Coppen, in a famous article on the biochemistry of affective disorders, that provided "one of the most cogent reasons for believing that there is a biochemical basis for depression or mania."[52] The physical methods—that is, psychotropic drugs—very quickly "evolved," according to Leo Hollister, a distinguished biomedical scientist who early worked on the pharmacology of the tranquilizers, "into tools for exploring the possible causes of mental disorders, as well as for treating them."[53] These "drug-induced" models of disorder, as he called them, brought the quiescent field of psychopharmacology to life in the 1950s, inaugurating a large multidisciplinary program of study into how the drugs act in the body and how their mode of action might be related to mechanisms of disease development and symptom reduction.

Some of the initial theorizing was reported in the popular press, with talk of how the tranquilizers "affect the globus pallidus" (1956 *Newsweek* article), the "production of serotonin" (1956 *Science Digest* article), and other brain structures and neurochemicals.[54] But the main work was quieter, in the scientific journals, where the lines of research expanded and intensified in the following years. For instance, early work on the drugs reserpine (one of the first antipsychotics) and iproniazid (the tuberculosis drug that became one of the first antidepressants) led to the initial hypothesis that low levels of monoamines such as serotonin and norepinephrine are significant to the onset of depression.[55] This hypothesis was enormously influential, energizing the whole field and eventually vaulting depression—understood as a rare episodic condition in the 1950s with full inter-episode recovery—to major importance in psychiatry and a central place in the DSM-III in 1980.[56] It helped stoke the hopes of the biological psychiatrists of moving their profession into the mainstream of medicine.

A number of developments followed in the 1960s and 1970s that contributed to a "revolutionary" change and a turn to a "biomedical orientation" in psychiatry.[57] The central issue at the heart of this revolution was diagnosis, an issue made urgent and inescapable by the belief in the relative specificity of drug action. "Until drugs appeared on the scene," Hollister wrote, "psychiatric diagnosis was a neglected art."[58] It was neglected, as noted earlier, because psychoanalysis identified a single root to all the neuroses—anxiety—and so did not depend very much on diagnosis or require strictly defined diagnostic categories. The individual case history was key, as it is for most forms of insight psychotherapy. In my interviews, for instance, the common way in which people distinguished their experience with a provider of therapy from a provider of medication was to contrast them in terms of their concern with diagnosis. The former wanted to talk about unique personal experience, while the latter wanted to identify symptoms that would make it possible to assign a specific diagnosis—and thus prescribe a corresponding medication. As this example already suggests, with the advent of the drugs, diagnosis became paramount and could no longer be neglected. The optimal and appropriate clinical use of the available drugs, thought at the time by scientists like Hollister to be "highly specific for particular illnesses," necessitated more precise diagnostic criteria and assessment instruments.[59]

The prescribing practices and advertising of the minor tranquilizers and other psychotropic drugs played an important role in spurring change. The wildly successful marketing of Miltown created a backlash, and as early as 1958, there was a congressional hearing on "false and misleading advertising of prescription tranquilizing drugs."[60] More agitation followed in subsequent years. Finally, the crisis that erupted over thalidomide—a sedative, which caused serious birth defects, taken by many pregnant women outside the US for morning sickness—led to the Kefauver-Harris Amendments to the Federal Food, Drug, and Cosmetic Act in 1962. The amendments gave the FDA new powers to regulate pharmaceuticals and their advertising. Among other things, the amendments required companies to limit the claims made in advertisements to those indications established before the FDA and to provide accurate information about side effects. In time, the FDA began to send out letters asking companies to cease advertisements that the agency found misleading and run corrective notices that explained the FDA objections.[61] For example, the FDA censured the maker of Serentil—an antipsychotic—for a 1970 ad that promoted the drug "for the anxiety that comes from not fitting in." The ad, according to the corrective notice ordered by the FDA and published in the same medical journal, "suggests unapproved uses" for "relatively minor or everyday anxiety," when in fact the drug is "limited in its use to certain disease states."[62] Drugs, under this new and subsequently tightening regime, were to be advertised, and by implication prescribed, only for the treatment of specific, medically defined and recognized *illnesses*, not the struggles of everyday life. And illnesses must be diagnosed.

Even more consequential for the issue of diagnosis was the amendments' addition of a new requirement for FDA approval. Beyond the existing safety standard, drug companies now had to scientifically demonstrate that drugs were "effective" in their intended use. The blind, controlled clinical trial, which would become a universal requirement for demonstrating drug efficacy by the 1970s,[63] was already beginning to be used with psychotropic drugs as early as 1955. But there were many obstacles to conducting rigorous trials by this method. It requires a group of study subjects who share the same diagnosis and can be randomly assigned to a treatment group (those who receive the drug being tested) and one or more kinds of control groups (e.g., those receiving a placebo or a different drug). But how could researchers get homogeneous samples of patients when diagnostic categories were poorly defined? "What one person may see as a 'schizophrenic,'" complained Hollister in 1975, "another may see as a 'depression' or a 'mania.'" This diagnostic inconsistency, he continued, "may militate against demonstrating the efficacy" of the relevant drugs and was a serious obstacle to progress in developing new ones.[64] Or, to give another example, how could researchers make accurate, blinded evaluations of outcomes without standardized, quantifiable criteria of assessment both before and after the intervention? In order to get beyond clinician "anecdotage," a method had to be developed that could assess the presence or absence of symptoms in some objective way.[65] The efficacy requirement, then, demanded more systematic and reliable diagnoses and methods of assessment.

Beyond advertising and measurement issues, drug "effects" also had direct implications for diagnostic practice and definition. If the action of a drug derives from its direct or indirect effects on a basic brain impairment of the person with a disorder—a deficiency of serotonin in the case of depression, say, or an excess of dopamine in the case of schizophrenia[66]—then another source of error that could confound appropriate prescribing was inconsistency between the action of the drug and the "disorder" it was being prescribed for. At the clinical level, Hollister proposed that patient responses to drugs might be used to reconsider and correct misdiagnoses, a practice that some doctors of the time seem to have been following. In a study published in 1973, for instance, psychiatrist Barry Blackwell reports evidence that diagnoses might be chosen or changed to match "the drugs available to treat them." He recounts a case history in which a woman's diagnosis is changed from "anxiety state" to depression after she responds favorably to a medication that is considered an "antidepressant." The clinician remarks after three months of treatment that "although she never looked depressed before, she looks less depressed now."[67]

This belief in the reciprocal relationship between specific disease and specific treatment is central to the medical model.[68] In the model, diseases are conceived as discrete malfunctions in the individual that arise from underlying pathological mechanisms. The validity of a diagnosis might be demonstrated, among other more direct ways such as biomarkers, from a predictable pattern of treatment responsiveness. Strictly speaking, if a drug is or approximates a "magic bullet"—the founder of chemotherapy Paul Ehrlich's famous term for medicines that go straight to and effectively attack the pathogens in the targeted cell structure while remaining harmless in healthy tissues—then it follows that a diagnosis might be inferred from the patient's response to the medicine.[69]

Moreover, it also follows that what is regarded as relevant symptoms of disorder might be defined in terms of drug effects.[70] By this logic, Hollister also argued that treatment response might be used to "tease apart" and refine the diagnostic entities themselves.[71] He contended, for instance, that "recurrent bouts of depression with little evident mania, should probably also be considered a manic-depressive variant, as some of these patients respond specifically to lithium carbonate."[72] Since lithium carbonate is typed an "antimanic" agent (its action is believed to target the mechanisms of mania), he suggested the positive response to the drug be used to change the boundary of manic-depressive disorder. With similar reasoning, researchers of schizophrenia argued that since studies showed that "drugs such as diazepam [Valium] and chlordiazepoxide [Librium] are more effective than phenothiazines [such as Thorazine] in relieving anxiety, one can conclude that anxiety per se is not a unique and primary feature of schizophrenia."[73] In this instance, the researchers argued for redrawing the boundaries of schizophrenia because the drugs effective for schizophrenia, the phenothiazines, apparently do not target the underlying mechanisms for anxiety, while the benzodiazepines, which are not effective for schizophrenia, do.[74] Throughout a growing research literature, definition and drug effect were in mutual interaction.[75] By the late 1960s, that literature had "mounted to thousands of articles."[76]

### The DSM-III

At least from the 1960s, drugs and diagnoses came to be bound up with one another in this close and reciprocal relationship. They have "'grown up' together," noted Allen Frances, who was deeply involved with both the DSM-III and DSM-IV (1994), referring to psychopharmacology and the "DSM system" after the DSM-II. The need created by the "psychopharmacological revolution" for better, more specific diagnoses was the "major impetus" for the structured assessments, diagnostic criteria sets and research diagnostic criteria (the so-called Feighner criteria),[77] and other tools that led to a fundamental revision of the *Diagnostic and Statistical Manual of Mental Disorders* (DSM-III) in 1980.[78] The DSM-I based the classification of disorders on models of underlying etiology, largely rooted in the psychodynamic and Meyerian psychosocial perspectives. The DSM-II was similar, though less theoretical, with room for more divergent viewpoints. The new DSM-III, by contrast, reflected its closely coupled relationship with research and psychopharmacology in a far more neurobiological orientation, with the drug action providing the technical means to materialize the theory.

The crucial goal for the DSM-III was a *reliable* system, one in which different clinicians would give the same diagnosis to the same patient and both clinicians and researchers would have a common language with standardized meanings. It carved up mental illnesses and everyday suffering into a large number of discrete, specific disorder categories (more than 250) based on descriptions of observable signs and symptoms that occurred together.[79] Anxiety, for instance, had been the "chief characteristic of the neuroses" in the DSM-II, with its Freudian/Meyerian theoretical framework.[80] Now, it was divided into a variety of separate disorders that greatly undermined its relative importance. The new categories were phobic disorders, anxiety states, and post-traumatic stress disorder (PTSD), each with several subtypes. Anxiety states, for instance, included panic disorder, GAD, and obsessive-compulsive disorder. Each subtype, in turn, had its own diagnostic criteria and inclusion rules. For instance, the three criteria for social phobia, a subtype of phobic disorder, were

> A. A persistent, irrational fear of, and compelling desire to avoid, a situation in which the individual is exposed to possible scrutiny by others and fears that he or she may act in a way that will be humiliating or embarrassing. B. Significant distress because of the disturbance and recognition by the individual that his or her fear is excessive or unreasonable. C. Not due to another mental disorder, such as Major Depression or Avoidant Personality Disorder.[81]

If the patient met the criteria, it could be *inferred* that he or she had social phobia; otherwise no diagnosis was to be made.

Further, it was critical for reliability (and the shift away from psychoanalytic theory) that the anxiety be overt and observable. In the DSM-II, anxiety might be "felt and expressed directly" or might be unobservable by being "controlled unconsciously and automatically."[82] But in the DSM-III, disorder symptoms are limited to instances in which "anxiety is experienced directly."[83] The requirement, consistent throughout the DSM-III, that symptoms be manifest and relatively unambiguous reflected the antipsychoanalytic view that only overt, immediately present symptoms reveal mental disorders. Reliability required overt presentation for two reasons. First, the signs and symptoms could not depend on a clinician's interpretation of internal dynamics or psychological processes. The whole point of the new edition was to create *standardized* criteria for research and clinical use, criteria that were clearly and precisely delineated and consistent across categories. In the absence of blood tests or other biomarkers, the diagnostic signs and symptoms had to be, and frequently were, "easily identifiable"[84] to any trained observer. Anything idiosyncratic, vague, or unique to a particular theoretical orientation would muddy the waters and prevent the manual from being the guide to reliable, measurable mental diagnoses that psychopharmacology and clinical prescribing had come to require.

Second, "easily identifiable" clinical features, requiring a low order of inference, necessarily limited symptoms to not only those that were directly observable by the clinician but to those reportable by the patient. In an interesting irony, despite the DSM-III's new commitment to the medical model, in which the disease not the patient tells the story,[85] the manual makes patients' own interpretations essential to most diagnoses. The key criteria for social phobia listed above (A and B), for instance, are based on the patient's own emotional report and personal evaluation of "normal." For patients to play this role in their own diagnoses, experiences of distress or disability must be transparent to them, regarded by them as inappropriate to their circumstances, and felt by them to be the immediate cause of impaired functioning in other aspects of their life.[86] They have to reframe private and personal experience as symptomatic, or at least a potential target of medical treatment. Only then can the clinician overlay that experience, which is only directly expressible in the vernacular language of everyday behaviors and emotions, with the DSM criteria. Thus, in an important sense, the DSM introduces and requires a new subjectivity, the self-managing patient who knows and can speak the language of symptoms. After 1980, as I'll elaborate in chapter 3, a vast array of psycho-educational efforts emerged to train this new patient.

Toward questions of "etiology or pathophysiological process," the official stance of the DSM-III was "atheoretical."[87] The manual was, to be sure, making a break with psychoanalysis and had a latent theory informed by a somatic orientation. But the new somaticists developed the manual in conflict with dynamic psychiatry, and in the late 1970s, their ascendency was far from complete.[88] The DSM-III was written and approved by the American Psychiatric Association amidst "heated controversy," as many dynamic psychiatrists sharply challenged the validity of categorical diagnoses based on symptoms.[89] In this conflict, the atheoretical stance split some differences, centering attention on "clinical manifestations" without pressing the issue of "how the disturbances come about."[90] Since the underlying etiology and pathogenesis of virtually all psychiatric conditions was unknown,[91] a system of diagnoses based on etiology was in fact not possible—including one based on biological etiology, the real gold standard for the somatic psychiatrists. Hence, the DSM, to quote from a nineteenth-century psychiatrist cited by Frances and his colleagues, was "forced to fall back upon the symptomatology of the disease."[92] Such a fallback position was not ideal. A later director of the National Institute of Mental Health (NIMH) would chide the DSM:

> In the rest of medicine, this [symptom-based diagnosis] would be equivalent to creating diagnostic systems based on the nature of chest pain or the quality of fever. Indeed, symptom-based diagnosis, once common in other areas of medicine, has been largely replaced in the past half century as we have understood that symptoms alone rarely indicate the best choice of treatment.[93]

But at the time, the new DSM-III was heralded as a milestone that placed psychiatry squarely in the medical model with a standardized, descriptive, and public language for diagnosing and treating mental disorders. The DSM architects hoped that it would advance research toward the development of more effective treatments—envisioned as mostly pharmacologic—and ultimately pave the way for a more valid diagnostic system based on etiology.

The medicalization of the field had been the DSM task force's goal from the beginning in 1973–74 when work began on a new edition.[94] Although research needs, clinical problems, and psychopharmacology were the major factors shaping the substantive changes to the manual, other pressures—institutional, financial, and professional—to create more reliable diagnoses also arose or intensified in the 1970s. Third-party payers, for instance, who would exert a growing influence over the financing of treatment after 1970, began to restrict reimbursement for mental health treatment, complaining of the lack of uniformity and clarity of psychiatric diagnoses and treatment modalities.[95] This reticence grew more pronounced as regulatory changes in the 1970s allowed first clinical psychologists and then social workers to bill for providing psychotherapy and modes of psychotherapy multiplied without clear standards of clinician accountability.

Psychiatry itself also came under an immense barrage of criticism in the 1970s for not operating within the medical model. This criticism came from many sources and had many dimensions. Since the 1960s, for example, the so-called antipsychiatrists, such as Thomas Szasz, R. D. Laing, and David Cooper, had carried on a popular polemic against psychiatry, accusing it of labeling and controlling not the sick but the unconventional.[96] Psychiatrists began to respond to this polemic in the 1970s. A similar emphasis on social control was also a prominent theme of feminism in the 1970s. In *The Feminine Mystique*, published in 1963, Betty Friedan attacked Freud and psychoanalysis for naturalizing women's subordination. Her book was followed by a steady stream of far more unsparing critiques of Freud and the scientific claims of psychiatry, beginning with Kate Millett's 1970 *Sexual Politics* and Phyllis Chesler's 1972 best-selling *Women and Madness*.[97] From a different tack, prominent scholars and popular writers in the 1970s attacked psychoanalysis and lampooned the eclectic psychological therapies, which had proliferated in the 1960s, for degenerating into *Psychobabble*, to quote the title of R. D. Rosen's popular exposé.[98] The scientist and Nobel prize winner Peter Medawar, for instance, characterized "doctrinaire psychoanalytic theory" in the pages of the *New York Review of Books* in 1975 as the "most stupendous intellectual confidence trick of the twentieth century."[99] Books such as Rosen's and Martin Gross's *Psychological Society* skewered popular psychological and therapeutic trends, from Freudian analysis to encounter groups, transactional analysis, "primal scream," "rebirthing," and much more.[100] Perhaps as damaging, a famous 1973 article in *Science* by Stanford psychologist David Rosenhan, titled "On Being Sane in Insane Places," publicly and prominently shamed psychiatry by seeming to show that psychiatric diagnoses were unreliable and essentially contextual.[101]

Other setbacks included a wider loss of confidence in the scientific validity and effectiveness of psychodynamic and psychosocial treatments. Concerns began to be raised in the 1950s, but new developments—such as the visible failures of the community health movement, launched in the 1960s as an alternative to institutionalization—cast new doubt on the role of psychodynamics in the treatment of serious mental illness.[102] Political agitation and dissatisfaction with psychosocial research, beginning in the late 1960s, created a funding crisis at the NIMH and led to growing demands that psychiatrists engage in "real research," according to stringent scientific standards of evidence.[103] The controlled outcome study used to test the efficacy of drugs became the standard for all treatments. The effectiveness of qualitative, relational, and time-consuming practices like psychoanalysis, which depended on the case history to document success—the "anecdotage" derided by Hollister—was hard to establish under these positivistic procedures; studies typically documented the short-term effectiveness of drugs but produced little systematic evidence for the benefits of "insight" psychotherapies.

Psychoanalytic and related therapies based their claims to legitimacy on a scientific expertise rooted in a trained capacity to distinguish between healthy and unhealthy with regard to matters of the mental and emotional life and professional competence with techniques to foster positive change. Just this scientific status was being challenged and increasingly rejected, as psychologists were themselves already beginning to have "grave doubts about the value of 'evidence' from introspection."[104] Under the new medical definition of good science, psychoanalysis, which had once represented reform, was now more often regarded as hidebound and an embarrassment.

### Psychiatry as Clinical Neuroscience

Along with the psychopharmacological revolution, these developments and others pushed psychiatry to move closer to medicine and the medical model and to enthusiastically and optimistically embrace somatic treatments and, if needed, short-term behavioral interventions (which also focus on symptom relief). After 1980, the field would move decisively in this direction, renewing a "militantly hopeful emphasis"[105] on understanding the role of biochemical and genetic factors so as to treat mental disorders. In the late 1980s, the introduction of fluoxetine (brand name Prozac) seemed to fulfill this hope and the promise of further breakthroughs. Even before Prozac, depression, once considered a psychotic condition, had undergone a progressively widening redefinition over the 1960s and 1970s—in part as antidepressants, such as Deprol (a meprobamate-antidepressant mix) and Elavil, began to be marketed for many of the same problems as the minor tranquilizers.[106] As noted earlier, depression was of intense interest to biological psychiatrists. In the 1960s, they had formulated a very influential hypothesis that monoamine depletion (a "chemical imbalance") was crucial to the generation of depressive states. This theory endowed "depression" with a special scientific status, and the somaticists elevated "major depression" to a correspondingly prominent place in the DSM-III.

Then, in the 1980s, scientists isolated fluoxetine, the first of the selective serotonin reuptake inhibitor (SSRI) class of "antidepressant" medications.[107] One version of the depletion theory centered on the monoamine serotonin. Fluoxetine was a compound that selectively targeted serotonin alone, suggesting the discovery of a "magic bullet" treatment. And so it was popularly received. Like Miltown and Valium before it, Prozac was a phenomenal public sensation, prescribed by general physicians, with blockbuster sales and relentless media attention and popular testimonials. But unlike the earlier drugs, Prozac was loudly touted not only for its promise of better living but for how it worked in the brain. Here, the public story went at the time, was a breakthrough discovery of neurochemistry, the first fruits of neuroscience to unravel the secrets of disorder.

In retrospect, however, the arrival of Prozac and the 1994 release of the

DSM-IV represented something of a nadir. In the years following, the validity of the DSM categories, along with the depletion hypothesis and other notions of "chemical imbalance" would come to be largely rejected in psychiatry. Yet, the biological turn had been made, and the NIMH and prominent psychiatrists would push further in this direction, calling the field to move beyond the heuristic categories of the DSM system. According to the first chapter of a leading psychiatry textbook, published in 2009, "there is little reason to believe that these [DSM] categories are valid."[108] Writing in 2013, Thomas Insel, then director of the NIMH, argued that "the strength of each of the editions of DSM has been 'reliability'—each edition has ensured that clinicians use the same terms in the same ways. The weakness is its lack of validity." Since, on this account, "mental disorders are biological disorders," then greater validity can only be achieved on the basis of a greater understanding of the biology of the brain.[109] The goal, according to the psychiatry textbook, is a "brain-based diagnostic system."[110]

No such system has arisen. When the DSM-5 was published in 2013, it made many changes but followed the same symptom-based format as the DSM-III and the DSM-IV after it. While much controversy surrounded the DSM-5, perhaps even more salient was the sense of disappointment. That letdown, the feeling that the field was stuck with a system it had superseded, comes across in the assessments of Insel and many other major figures in the field. According to a common view, expressed in an editorial in the prestigious *JAMA Psychiatry* in 2015, psychiatry is coming to recognize that its future is in clinical neuroscience.[111] The DSM-III revolution was based on the belief that psychiatry, especially through psychopharmacology, had begun to penetrate the biology of mental disorder. Today, however, the editorial insists, "new findings have reshaped the fundamental way in which we understand psychiatric illness." Tellingly, the authors use the example of depression, so critical to the DSM-III, whose etiology "was once characterized as simply a monoaminergic deficit." Now the field has moved beyond chemical imbalance theories, and "new research is expanding our understanding of depression across multiple levels of analysis—from circuits, to neurotransmitters, to synaptic plasticity, to second messenger systems, to epigenetic and genetic differences." All of this, the authors assert, should be leading to a clinical "paradigm shift" that is in fact not happening. Instead, barriers are preventing the field from "embracing a new identity." The biggest barrier, they lament, is the "pervasive belief that neuroscience is not relevant to patient care." They then grudgingly acknowledge that in fact all the research "has not translated into routine clinical interventions."[112] And, they might have added, the scientific challenges to understanding are growing not receding.[113] Despite the militant hope, the effort to identify biomarkers, according to other researchers, has brought "three decades of consistently negative results."[114] As was the case in 1980, no knowledge foundation exists on which to build a system of diagnoses based on etiology.

Without an actual alternative, the symptom-based DSM remains in full force. And it can't be abandoned, even if, as Insel and others maintain, it lacks validity. As already suggested above, the rationalized systems of medicine that have arisen since the 1970s require diagnostic categories within a specific disease model. Former director of the NIMH Steven Hyman provides a summary of the many functions that diagnosis now has in psychiatry that is worth quoting at length:

> Reliable and widely shared disease definitions are a necessary antecedent for rational treatment decisions. Diagnosis guides a clinician's thinking about treatment, about other symptoms that might be present, about likely impairments, and about prognosis. Shared and reliable diagnoses are the cornerstone of communication between the clinician and the patient and, where appropriate, with families, other caregivers, and institutions. Diagnosis is also central to translational and clinical research: Without clear diagnostic guideposts, idiosyncratic groupings confound clinical trials, epidemiology, genetics, imaging, and other laboratory studies. In their absence, academia and industry lack indications for which to develop new treatments, and regulatory agencies cannot judge efficacy. Diagnosis also plays an important role outside the clinic and laboratory, influencing, for example, insurance reimbursement, determinations of disability, school-based interventions for symptomatic children, and diverse legal proceedings.[115]

All of these clinical, scientific, commercial, and administrative functions have deeply embedded the DSM diagnostic model within the bureaucratic delivery of health care, as well as most of the federal agencies who fund research and regulate drugs, devices, marketing, and more. After 1980, seeking help for behavioral and emotional problems in almost every health care setting meant talking about problems within the terms of the DSM. And this institutionalization in turn carried DSM categories into virtually every context in which personal problems or disruptive behavior might be at issue, whether the school, the church, the courts, the military, human service agencies, counselors' offices, human resources departments, and so on. Over the past three decades DSM categories like depression, PTSD, social phobia (now, social anxiety disorder), and ADHD have become part of the very institutional fabric of contemporary society.

Besides shaping language and understandings, institutionalization had another effect. It gave the DSM categories a powerful facticity and reinforced the turn to biology. In precise terms, the categories are heuristics, useful tools for certain purposes; but in practice, as Insel, Hyman, and many others have noted, they have been widely confused by both professionals and the lay public with real entities in nature, independent of human observers. Given their use throughout medical and other service and commercial contexts, it is easy to see how this would happen. In all these contexts, psychiatric diagnoses are made functionally equivalent to other medical diagnoses. Where on the reimbursement form or medical record, for instance, does it distinguish between disorders that are valid and those that are merely reliable? In what drug trial report or epidemiological study do the authors qualify their assertions about findings by noting that those they group together may not be suffering from the same thing? The whole medical-scientific complex, clinical and research, effectively demands reified categories. In institutional practice, all medical diagnoses label disorders, illnesses, or diseases, all give a name to something a patient suffers from, and all are objects of physician intervention, quantification, and research study. For practical purposes, all are the same kind of thing, given, at least in everyday institutional contexts, the same ontology. No wonder the confusion.

Further, while psychiatric best practice guidelines would continue to put medication and psychotherapy together as they once did, actual medical practice moved decidedly in a biomedical direction. As noted earlier, there was a high rate of psychotropic drug use in the tranquilizer era, at more than 20 percent of the adult population. Those rates would decline in the late 1970s—especially as worries about addiction led to a sharp decline in Valium use—and then accelerate again after 1990. The rates are sky high again.[116] But there are many differences in contemporary practice, differences that reflect the shift toward a biomedical orientation. Four trends stand out.

First, patients take popular medications far longer than they used to. The antidepressants, for instance, have come to be used at nearly the rate that Valium was (13 percent of adults now, 15 percent then).[117] Reflecting the notion of a chronic brain abnormality, most patients who take an antidepressant, unlike Valium in the earlier era, will remain on it for years.[118] Second, there has been an enormous increase in the number of patients who take more than one psychotropic drug (from the same or different classes) at a time.[119] This practice, known as psychotropic medication polypharmacy, reflects the narrow "target-symptom approach"[120] of biological psychiatry and is now very widespread.

Third, and perhaps the practice that most illuminates the shift toward a biomedical orientation, is the practice of prescribing psychoactive drugs for children. While the prescription rates for adults were high in the 1960s and 1970s, the numbers for children were vastly smaller. Based on prescription data as late as 1984, researchers could observe, "As a class, psychotherapeutic drugs are not often used in people less than 20 years old."[121] In the intervening years, the situation has changed profoundly, with an exponential increase of prescriptions to children as young as preschoolers and to adolescents. In 2014, researchers were observing that "mental health problems are common chronic conditions in children" and "medication is often prescribed to treat the symptoms of these conditions."[122] This new, even enthusiastic willingness to medicate reflects the view that the problems are neurobiological, constitute a risk for greater problems in the future, and should be treated by drugs because they address (but do not cure) the underlying abnormality.

A final revealing aspect of the turn toward a biomedical orientation is in the sharp move away from psychotherapy. Insight psychotherapies, which had flourished alongside rising rates of Valium use, have been progressively abandoned by psychiatry in favor of pharmacotherapy. Psychiatrists now see patients for med checks, and when they do recognize the need for psychosocial interventions, they often farm this task out to others (a practice known as the "split-treatment" model).[123]

In a crowning irony—and one of the clearest signs of how far the biomedical orientation has triumphed—even psychotherapists have begun talking about how talk therapy works by unleashing "highly specific biological changes in the brain," such as affecting "cerebral metabolic rates" and "serotonin metabolism."[124] No wonder psychologists have been so anxious to secure at least limited medication prescription privileges. The goals of talk therapy and medication treatment have become entangled, and both, it seems, are to be understood in the language of the brain.[125]

## Conclusion

The post-1970 changes I have described, including new drug marketing regulations, the decline of Freudian and psychosocial perspectives, the DSM-III, and the biomedical turn of psychiatry, fundamentally changed the healthscape and the terms in which doctors and other medical professionals talk about and treat cognitive, emotional, and behavioral problems.

Gone is the earlier and more vernacular talk that accompanied the first great wave of popular (prescription-only) drug-taking. In the tranquilizer era, despite widespread medication use, the distinction between serious psychiatric conditions and everyday suffering was de facto preserved. But in the years after 1980, the vernacular was steadily replaced by an extensive diagnostic language of mental disorders. The new language did not, by definition, name the troubles that the minor tranquilizers were once marketed for, such as a relationship bind, trouble adapting to distressing circumstances, or an unfulfilling life situation. It named something you *have*, an internal dysfunction. While many psychiatric theorists now emphasize that mental disorders involve a complex interplay between biological and psychosocial factors, they also insist that all disorders are ultimately neurobiological. Even "the study of unconscious processes, motivation, or defenses," according to one influential paper, "is now also in the domain of cognitive neuroscience."[126]

Psychiatry and medicine more generally now find themselves without any clear neurobiological basis for clinical work and, at the same time, without depth models of subjectivity and the sort of languages for exploring inwardness that approaches like psychoanalysis, phenomenology, and spiritual traditions once offered. It is this new biomedical healthscape, with its neuro veneer, that helps us to situate the features of the clinical encounter that Eric, who had the "meltdown" while teaching at a competitive school, reports. Predictably, his physician made a diagnosis based on the DSM and explained the role of medication with the notion of a chemical imbalance. She also downplayed the ameliorative potential of other calming interventions. Apparently, any thought to exploring the meaning of Eric's anxiety simply never arose.

Meanwhile, even before the appointment, Eric knew what the diagnosis would be and came prepared to get a prescription. He had already been exposed to the biomedical healthscape, which has many different and interacting sources beyond the clinic. These sources include discussion with family and friends and a wide variety of popular media, such as drug advertising, magazine and newspaper stories, autobiographies, public information campaigns, websites, social media, and more. For ordinary people, these sources are where they typically hear about disorder categories, symptoms, medications, drug effects, brain states, and the like. We cannot understand the neurobiological imaginary without considering this pervasive talk and the explanations and promises on offer.

# THREE

# Appropriating Disorder

About five years ago, Georgia, now age forty, lost her job. At first, she thought of it as an opportunity and decided to go back to college part-time to finish her degree. But then financial adversity struck. She remained unemployed for a year, and not long after taking a new administrative position, her husband was downsized out of a job in a company restructuring. Georgia was, suddenly, solely responsible for their joint finances, forcing her to cut back on classes and make significant lifestyle changes. She found these events overwhelming to the point, she says, that she was "just not myself." She was often sad, couldn't motivate herself to do anything productive, and felt that she was at a "standstill" in her life. She began to question herself and reevaluate her choices. And though she knew it was unfair to blame him, she found herself feeling resentful toward her husband for losing his job. She felt she was "being cheated" because of his situation.

Georgia began to think she might be suffering from depression. She had read magazine articles and seen pharmaceutical advertising that presented different symptoms of depression. She found herself identifying with some of the listed "items." About one of these encounters, she recalls:

> I think I was looking at something on TV or reading something in a magazine and two of the symptoms kind of related to me, and I said, "Well, this is probably what I'm experiencing." . . . I'm reading these things and seeing these items and it's like right there. I'm relating to them right there.

And that, she said, "kind of triggered it." She knew she had depression. Georgia also turned to a close friend who had lost her husband, as well as a sibling, and had been diagnosed with depression and was taking medication. When her friend suggested Georgia see a doctor, she quickly followed through.

Given her "self-diagnosis," Georgia wasn't surprised when the physician diagnosed her with depression. The doctor, as she put it, "kind of concurred with me." But even if expected, she found it hard to "face the reality" that she suffered from depression and was very reassured by the way that her doctor described the condition. The doctor emphasized that being depressed does not mean a person is crazy or suicidal, that it can happen to anybody at some point in life, and that it is nothing to be ashamed of. "She said it was normal," Georgia reports, "and so that made me feel good."

Even if her depression is "normal" or "normal to a degree," Georgia *has* something, which, she says, "can be a form of disease." Her depression is not, she emphasizes, the result of a brain disorder but has come on in response to setbacks in her life. That means to her that she does not have a mental illness and has not been "diagnosed with mental illness." Rather, "it's depression." But this depression is like an illness in the sense that it arises from outside herself. And it can get worse to the point that it overwhelms you and makes you "feel just totally lost." According to Georgia, she is the kind of person who is confident, takes control, and addresses problems before they get out of hand. Her experience is anomalous because tough circumstances alone would not cause her to become so sad and unmotivated. Some sort of medical condition—a "normal depression"—has intervened.

Georgia and her doctor did not discuss counseling, and, Georgia indicates, she never gave any thought to seeking psychotherapy. She has a medical issue and had seen the drug ads, read the magazine articles, and discussed all this with her friend. When the doctor recommended an antidepressant, that too came as no surprise. After first doing some internet research on the side effects of the prescribed drug, she decided to take it.

Like her friend, Georgia found the drug helpful, though the effects were quite subtle: "When I first started taking it there was no major magic wand that made me like, 'Oh, wow. Okay, I'm not depressed anymore.'" But she began sleeping better and could get things accomplished at work and around the house. She was calmer, less emotional—changes that increased her confidence and self-esteem. These gradual improvements, combined with a better financial outlook and the chance of a promotion at work, have led Georgia to feel more in control, more herself again. She had been on the antidepressant for eight months at the time of the interview and is wondering if she still needs it. She is sure that she doesn't plan to be on it all her life.

------

As we saw in chapter 2, cognitive, emotional, and behavioral problems of every conceivable variety are now discussed in medical and other institutional settings in a diagnostic language of mental disorders. This language has an implicit biomedical (somatic) orientation and has become tightly linked in psychiatric practice with the prescribing of psychotropic medications to the neglect of other avenues of treatment (with the partial exception of short-term, symptom-focused behavioral therapies). The new biomedical orientation presupposed and required a new type of psychiatric patient and was never merely a professional phenomenon; it circulated through feedback loops with popular representations and interpretations. Georgia's story is an interesting illustration of how the new language and assumptions might be acquired and incorporated within a narrative of self. It shows, even in the absence of direct brain talk, the neurobiological imaginary at work.

According to Georgia, she basically diagnosed herself with depression, drawing on representations of symptoms she had already encountered in popular media, advertising, and the input of her friend. Her understanding does not actually comport with the DSM concept; in the DSM, normal and expectable responses to life events are excluded from its definition of mental disorder, and depression is a type of mental illness.[1] But Georgia believes her understanding is shared by her doctor, was even suggested by her doctor. In their interaction, they were, in fact, using the language of depression. They agreed that what she has been experiencing *is* depression, an idea she embraces as a clinical truth even as she animates it (in effect though not necessarily in intention) with her own meanings.

At the same time, Georgia appears to reject any notion that she has a neurobiological disorder. She does not attribute her depression to a malfunctioning brain. Yet, it is nonetheless presented as the effective cause of her suffering. Her depression, whose reality was hard to face and which she has kept carefully concealed even from her husband, is actually an alien force. It is not her own response nor an indicator of anything about her or her psychology. Hence, counseling or therapy were never indicated. Rather, medication was the only appropriate response. With it, she can "take control" once again.

In dealing with her predicament, Georgia forged a link with depression (and then medication) as her experiential knowledge spontaneously developed in interaction with assorted mediators of the new biomedical health-scape, including the physician, magazine articles, direct-to-consumer advertising, and a close friend. What variously circulates through these sources are not the abstractions of research psychiatry or even simple translations of professional concepts and science into more popular terms. Rather, each functions as a forum and a vehicle for practical interpretations of symptoms, applications of DSM categories, and advice about the appropriate actions to take. And for laypersons, the personally meaningful engagement with such sources comes in the immediate and real-world context of predicaments and a search for answers.

In this chapter, I explore common sites of the healthscape, places *where* participants encountered biomedical language and ideas. I investigate *what* explanations and promises they heard or might have heard there, and I consider *how* they engaged with both. The grammar of the neurobiological imaginary is something acquired; it displaces a vernacular episteme through encounters in which suffering comes to be grasped in new and objectifying languages and practices.

## Personal Appropriation

In their help-seeking, participants, like Georgia, report contact with a variety of sources of information and feedback. These sources, whether a website, a doctor, or a close friend, served as interlocutors. They were engaged to help the participant think about his or her predicament and what to do. Of course, for many, the initial turn to family or friends or a professional was in search of personal support and assistance. In the great majority of cases, they also received, and this includes their contact with popular media, something else. Each source encouraged or helped to facilitate a personal identification with a DSM category.

Some of those who identified with a category, like Georgia, spoke of a "self-diagnosis." This term, though, is misleading. Only a medical professional can make a diagnosis, and personal identification could come before, after, or even in the absence of such a professional decision. What this indicates is that in practice, diagnostic categories have two meanings. The first I will call, following several of the people interviewed, the "formal" diagnosis. This is the diagnosis given by a qualified professional and having whatever meaning the clinician might have provided (I have no direct evidence of that meaning, of course, but participants who had any such discussion with a doctor generally took it to be a canonical medical view).

The other meaning of a diagnostic category is the meaning that the participant gave it. In the interviews, we were careful to explore this personal, individual meaning. I had observed that interview studies typically recruited diagnosed patients and then queried them about their diagnosed condition without asking if the participant actually accepted the diagnosis, used the language of that condition in accounting for their experience, or, if they did, what they meant by it. As generally in the medical research literature, there is an assumption that terms like depression or social anxiety or attention deficit have a stable meaning across the social contexts in which they are employed. Against that assumption, my strategy was to let people tell us what language they used and how they used it.[2]

As with Georgia, the personal definition was very self-referential, a definition in terms of "what depression (or social anxiety or attention deficit) means to me in my experience." This improvised meaning is another reason why the notion of "self-diagnosis" misses the mark. People, as I interpret their reflections, do not passively accept DSM categories like depression as predefined and natural kinds of illness, on the model of diabetes or heart disease.[3] Rather, they apprehended or annexed diagnostic language, fitting it to their experience through a felt identification, as they sought to explain their suffering and pursue an ordering of self in light of normative demands. Rather than self-diagnosis, I will speak of this active, responsive engagement as personal appropriation.[4]

The appropriation comes in the process of making sense of experience and "purifying" it of undesirable connotations. According to Georgia, thinking of depression as a normal response and not a mental illness was crucial to completing the medical connection she describes as "triggered" by symptom presentations in magazines and ads. In this idea, she found an "impression point" that worked to transpose the explanatory power of medical concepts into a new understanding of her experience and herself. As theorized by the anthropologist Peter Stromberg, the impression point is a process with three interrelated features: "a new understanding of self, a new understanding of a symbol system, and a feeling of commitment," all generated together.[5] These features can be illustrated in Georgia's experience: a perception, an image of herself, is formed as the familiar category of depression acquires a new and unsuspected significance; the formation of that image coalesces through "an infusion of subjectivity" wherein she animates the category with her own meanings; and the image is built up from this discovery point, the finding of an important part of herself in the medical symbolism, generating commitment and emotional investment in the symbolism. In this cultural model of symbolic appropriation, person and symbols are in a dialectical relationship.[6] Georgia makes a commitment to "depression" which re-forms her by being taken up into a revised understanding of herself and her suffering.

By such a process, two-thirds of participants appropriated a disorder category. The other third did not (in all but six cases, they had never seen a doctor or been formally diagnosed). This latter group mentioned contact with the same public sources, as well as friends and family. They too were well aware of and had to contend with the healthscape of biomedical language and pharmaceuticals, even when they had never seen a professional.[7] So, although they rejected or passed on a diagnostic category and did not pursue medication (with one short-lived exception), they were involved in an interpretive process quite like those who did.

In most cases, people regarded a category as *in*appropriate because they adjudged their experience to be of a different kind than a psychiatric disorder or mental illness. This is noteworthy because they too saw their suffering as abnormal, and there is no obvious difference in the predicaments experienced by those who appropriated a category and those who did not. Tewanda, for example, age forty-seven, was working temporary assignments while looking for a full-time position. She was facing very difficult circumstances at home and describes herself as feeling "sad" and "lost" almost all of the time. But she does not consider her experience the same as depression. Depression would be more like what her mother experienced after her father died. While her mother used to be "a lively person, kind of outgoing," she is now listless and fretful, dwells on the past, is easily discouraged, and is more dependent. In Tewanda's view, something is wrong with her mother. Her own experience, by contrast, is primarily due to her circumstances, and though struggling a lot, she has some control and believes that "one day I'm going to get through all this without taking medication."

Tewanda rejects a medicalizing explanation based on her judgment of what constitutes a disorder, what treatment it might require, and what either or both would indicate about herself and her personal control. These were the same issues that both those who appropriated categories and those who did not confronted. Accounting for experience in medical language raises the question of mental disorder or illness—typically conceived, explicitly or implicitly, by participants as a somatic malfunction that has some direct determinative effect on one's thoughts, emotions, or actions— and the implications of that idea for treatment and self-understanding. Georgia found an impression point precisely in an understanding of "depression" that was stripped of "mental illness." Tewanda, who had a different idea of "depression," found no such connection. She could not locate herself within the same category as her mother. Georgia, and others like her, *could* appropriate because, in many cases, they knew diagnosed persons with whom they could identify; and, as I'll elaborate more in the next chapter, they established a distance between the diagnostic category as appropriated and undesirable connotations of mental illness.

In considering the roles that the common interlocutors play in help-seeking and fostering appropriation, I begin with clinicians.

## Interaction with Medical Professionals

In Georgia's account of her experience, the physician confirmed her distress under the diagnosis of depression, and in this context apparently reassured Georgia by naturalizing her feelings as common and expectable responses. The understanding of depression that Georgia took from popular media and interaction with the physician is not a medical one, and it is doubtful if her doctor would independently endorse it as the meaning of major depressive disorder. But it is what Georgia *heard*, and in her account, the doctor concurred with her judgment. The practical matter at hand was what to do about it.

When participants referenced clinicians, they mainly spoke of their mundane role in the (public) decision-making process. The pathway to the professional was triggered by a wide range of experiences and took a variety of routes, sometimes passing through a nonmedical therapist. Family or friends were often critical intermediaries, as I'll return to below, and for at least five participants (one was unclear), a diagnosis of a mental disorder arose in the context of a medical visit for a different reason, such as a routine physical exam or to discuss physical symptoms. When asked to describe what transpired in their initial interactions with the doctor, most people, whether they had seen a psychiatrist or a general practitioner, provided little more than a chronology (and a few could not remember any details). They talked, and the physician listened, asked questions, and perhaps took them through a diagnostic questionnaire. The doctor then made a diagnosis and recommended treatment.

In these stories, people had little to say about medical professionals as direct influences on their (personal) explanatory perspective. There were exceptions, like Eric from chapter 2, who cited his physician's persuasion as leading him to a medicalizing perspective. More commonly, depictions of the physician role were of two general kinds. In one, participants portrayed themselves as identifying with a diagnostic category, and often coming to a medicalizing view and deciding they needed medication, largely independent of a physician. In most of these cases, like Georgia, they reported that once they saw a professional, he or she agreed with their evaluation and prescribed readily. In other cases, they portrayed the professional as more hesitant, yet they still came away with a prescription. Recall Kristin, the college student whose story I told in the introduction. Her doctor told her that she did not have all of the symptoms for an ADHD diagnosis yet prescribed a medication nonetheless. In a few additional cases, the professional refused, though in these instances, the participant pursued a different professional or professionals and ultimately got a prescription. In all these cases, people describe the diagnosis and prescription as (ultimately) confirming their category appropriation and explanatory understanding. The professional did not introduce, or they did not accept, anything different.[8]

In the other common pattern, participants report some initial resistance on their part to a diagnosis, a medicalizing explanation, or a prescription. They indicate that their doctor, often during the first appointment, made a diagnosis, recommended medication, and introduced the idea that their problem had a neurobiological cause.[9] Some found this surprising and a reason for concern. In our interview, they complained about how quickly and casually the physician or psychiatrist made a diagnosis and wrote a prescription, and they described the interaction as impersonal and narrowly focused on symptoms. After college, Byron, for example, began working in a job that required extensive travel. Grappling with the physical and mental demands of his work, being apart from his girlfriend, and struggling with questions about his future, he felt worried, overwhelmed, and confused. At the encouragement of a friend, he saw a psychiatrist who diagnosed him with a "common stress disorder" and prescribed a sedative. Byron took the drug briefly but then quit, critical of the doctor for making a snap judgment.

In other cases like this, participants reported that their own views on diagnostic category and causation only took shape later, based on their experience with medication or other sources of information. Several specifically contrasted their personal understanding of their diagnosis with what they took to be the doctor's professional view.[10] But in virtually none of these cases did people report actually discussing their views with the physician. They present the interaction with the physician as formal and one-sided, with their questions of the meaning of diagnosis and treatment engaged outside the doctor's office.

However, across both patterns there was one way in which medical professionals were critically important to the interpretive process, especially but not only for those who proceeded to medication. They were a crucial "witness" to the rightness of the participants' help-seeking and medication-taking, and their diagnosis gave meaning to and conferred a reality and legitimacy on the participant's suffering. People frequently cited the fact of a formal diagnosis as important in how they accounted for their experience to others and to themselves. The very act of official naming gave anomalous experience a kind of stability, separating it from the flux, imposing a socially recognized meaning—it is *this*, and not *that*—and linking it to a body of expert knowledge. Richard, for example, a business consultant, succinctly described this organizing power. He reported a history of work-related "problems in . . . a certain level of emotional intelligence that it takes to be in a cooperative relationship with people." He thought he was depressed and was surprised to receive a formal diagnosis of ADD at the age of sixty-one. He found it "sort of an epiphany in the sense that it helped to explain some setbacks I had in the course of my working life."

The formal diagnosis has its own symbolic power. Participants affirmed it even when they self-consciously infused the diagnosis with their own meaning, and even when they emphasized that the doctor only regarded them as partially meeting diagnostic criteria. They also affirmed it when they worried that the diagnosis was made too quickly, without sufficient care. For many people, this symbolic power was critical to their efforts to find meaning and direction. On anomalous and ambiguous personal experience, it conferred a sense of order and objectivity.

Of course, I am describing what participants said about their interaction with psychiatrists and other clinicians, and I have no independent knowledge of what was actually said or done. There is, however, much independent evidence in support of participant observations about their active role and appropriation of diagnostic categories and the casual and cooperative diagnosing of physicians.[11] Medical practice itself has come to reject paternalism and encourages, even requires, active patients.[12] As we saw in the last chapter, the DSM builds patient knowledge and evaluations into the very criteria of many disorders. Interview participants themselves made it clear that they felt the responsibility to be engaged. With only a few exceptions, they saw it as their prerogative, even their obligation, to define their situation. When professionals offered interpretive input, it seems to have been regarded as advisory and only accepted when it concurred with their own assessments. As one participant, Ernest, forty-seven, stressed, no one has the right to define his condition.

While the clinician's role in participants' stories was often limited to the decision-making context, popular sources stood out as interpretive guides. Appropriating disorder and accepting medication required help, help that participants needed both prior to and after seeing a professional. This process too was, in crucial respects, dialogical.

## Popular Sources

The move to a biomedical orientation in psychiatry was an effort to place the field on a more secure medical footing. The decisive step was to move away from the seemingly imprecise and poorly defined world of everyday suffering and mark out the specialty in terms of specific disease entities and specific treatments. Compared with dynamic psychiatry and the prescribing of the minor tranquilizers, the central place of diagnosis represented a new and higher standard, seemingly limiting the scope of concern to the treatment of people with reliably classified disorders.[13] Further, as disorders increasingly came to be defined as problems of the brain, the direction of psychiatric treatment could have been expected to narrow and move toward more debilitating forms of mental pathology. The many kinds of circumstantial dilemmas and emotionally stressful experiences treated with drugs like Valium in the tranquilizer era would seem to have been left out. Not because these experiences might not cause pain or hardship or even come to medical attention but because no pathology was present, and pathology was the appropriate object of medical treatment.[14]

Perhaps for a time they were left out. Prescription rates for many common psychotropic medications, for instance, did in fact drop in the 1980s. Of course, many other factors were at work, but one thing is clear, as noted in chapter 2: the DSM-III, even while trying to contain the meaning of diagnostic language, gave sufferers a new and important role in the diagnostic process. For many of the most common categories, the criteria required that they think of the emotional and behavioral problems they might struggle with in terms of "symptoms" (and, thus, at least implicitly, in terms of illness), regard them as inappropriate to their circumstances, and construe them as an interference with other important aspects of their life. Some of this role may be similar to what it was in the tranquilizer era, reflecting in part the common general process by which ordinary people make judgments about whether they have a mental health problem and might need help. These judgments typically reflect the sense that one's experience is unusual, unreasonable, or difficult to comprehend and attributable, at least in part, to an aspect of one's internal makeup.

On the other hand, the DSM expectations of sufferers were also new and specific and required both knowledge and interpretation. There is no clearer evidence of the role for this new type of patient than the vast array of psycho-educational efforts that emerged since the DSM-III to publicize the new or renamed categories (there are more than 250 categories in the DSM-III and even more in the DSM-IV and DSM-5) and their symptoms. Popular media, from magazines to the internet, provide extensive mental health news and commentary. Newspapers, for instance, added health sections in the early 1980s, in part to address the perceived need for more mental health coverage.[15] Since the 1980s, patient advocacy and medical identity groups have been steadily growing in number and importance.[16] These groups lobby Congress; conduct national "screening days" for specific disorders; produce public service announcements (PSAs); maintain websites with information, symptom checklists, treatment options, and testimonials; sponsor popular books—*Triumph over Fear* and *Triumph over Shyness: Conquering Shyness and Social Anxiety*, to give two examples from the Anxiety Disorders Association of America[17]—and engage in many other educational activities of this kind in order to foster help-seeking.

Besides funding research, federal agencies such as the NIMH, the Centers for Disease Control and Prevention, and the Office of the Surgeon General also issue public reports on mental illness, produce PSAs, and have extensive public information websites that list "warning signs" of disorders, their symptom profiles, and available treatments. Professional associations, such as the American Psychiatric Association, provide many of the same types of educational information, as do, in various ways, patient blogs, self-help books of a wide variety (e.g., *The Hidden Face of Shyness: Understanding and Overcoming Social Anxiety*, *The Anxiety and Phobia Workbook*, and countless others),[18] best-selling autobiographies (*Prozac Diary*, *My Age of Anxiety*, *The Noonday Demon*), and much more.[19] All this, plus the phenomenal explosion of direct-to-consumer advertising since the late 1990s.

Everyone has been exposed to this biomedical healthscape. In the interviews, we asked participants about what popular sources they had encountered or consulted and their thoughts on what they had learned. All had come upon many sources of ideas and images. Unsurprisingly, they could usually not separate out the discrete messages they had heard or been influenced by. Rather, what they attested to was a general familiarity with common messages. This familiarity came across, for instance, in the easy use of drug names and diagnostic categories, in the casual references people made to the workings of the brain and neurotransmitters, and in their talk of the brain as a kind of independent actor. It came across in the lines they used to mark off "deviant" (nonnormative) behavior or emotion or unacceptable limitations, in the high expectations for drug effectiveness, and in the ready acceptance of diagnostic categories as valid biophysical conditions. All such talk and much more demonstrated the wide social circulation of biomedical ideas and assumptions. Not everyone embraced the messages or shared the assumptions, of course—far from it. But the point is that the sheer ubiquity of talk about symptoms, disorders, and drugs has produced a common pool of "knowledge"—part of the healthscape—that participants encountered, contended with, and drew upon in evaluating and accounting for their everyday suffering.

If this common knowledge is the iceberg, we get some notion of its tip from participants who commented on the direct influence of media, pharmaceutical advertising, and friends and family members who offered advice, personal experience, referrals, and, in a few cases, drug samples. I will consider each briefly, exploring how they encourage and facilitate an impression point and personal appropriation.

### Popular Media

While a number of interview participants mentioned popular media sources, including self-help books, magazine articles, and, especially, internet sites and social media, they spoke in generalities and had little specific memory of what they learned. As with Georgia, it was sometimes hard to distinguish between exposure to popular media and advertising, particularly when the participant reported contact through social media or in conducting internet research into a particular diagnostic category or a branded medication. However, this may be a distinction without a difference. Many websites are produced or sponsored by pharmaceutical companies, and representative studies of mass media show that articles depict mental health in highly medicalized terms and portray common forms of everyday struggle and suffering as symptoms of disorder. In both respects, this is very similar to direct-to-consumer pharmaceutical advertising. Among many such studies, two will suffice to show this trend.

A first media study, published in 2009, provides an informative example of the growth of the biomedical healthscape and the lay trend toward the neurobiological imaginary. The study explored the depiction of depression in high-circulation magazines, such as *Forbes*, *Newsweek*, and *Redbook*, over the twenty-five-year period from the DSM-III in 1980 to 2005. The researchers coded their analysis of the articles in terms of how each addressed questions of definition, causation, and treatment. They also noted what was *not* said. The overall trend, the researchers found, showed a sharply intensified biomedical understanding of depression over the time period, an increasingly positive depiction of drugs, and a striking growth in the endorsement of medication as the preferred treatment. The reporting on drugs in the articles included almost no discussion of their side effects or the length of time the medications needed to be taken.

In the 1980s, the study found, depression was described in many ways, some more narrowly medical and some encompassing common problems related to personal and social changes and everyday suffering. But by the 1990s and beyond, the definition, according to the study authors, "almost entirely relied on different aspects of biology, biochemistry, genetics and other . . . explanations from the human biological sciences."[20] Similarly, while there was some "biologization" of depression in the 1980s, the putative causes were mostly normal life transitions and social experiences. In the 1990s and after, there is a dramatic narrowing of the causes, and these become almost exclusively identified with malfunctioning in the body/brain. The perception of the incidence of depression also rose after 1990, along with the notion that anyone and everyone might be susceptible. Finally, the study found that magazine articles in the 1980s suggested many possible responses to depression, including therapy, social support, and exercise. In the 1990s and after, by contrast, the focus is on getting expert help, and the solution is "almost unilaterally drugs" and "sometimes psychotherapy as an adjunct to medication."[21] By this point in the book, that summary will sound entirely familiar.

Further, as I have stressed, the relationship between lay representations and psychiatric discourse is not one of simple translation or reproduction. Much of what transpires in popular media does not mesh with or reproduce DSM criteria. Consider a second study of popular magazine and newspaper articles covering a roughly similar time period (1985–2000). This study compared the characterization of depression in the articles with the DSM language of depression. Since the advent of the new SSRI antidepressants in the late 1980s, it found a widening set of non-DSM criteria for defining "depression."[22] With respect to women, DSM-derived terms were progressively replaced with problems of marriage, motherhood, and menstruation. The studies' comparison case, which followed articles about antihistamines over the same time period, showed no deviation from medical criteria (as specified in *Cecil's Textbook of Medicine*).

The availability and use of the SSRIs, in other words, is temporally associated with a change in popular representations of depression, an enlargement of the category to include more common life problems and everyday suffering—even as the condition is explained in increasingly somatic terms. We have encountered this enlargement before. In the last chapter, I noted the magazine articles that began to appear in the 1950s touting the effectiveness of the new tranquilizers for treating everything from fatigue to lack of "social ease," performance problems, and much more. The difference is that those articles often made little reference to specific medical conditions, while in our post–DSM III and far more bureaucratic world of medicine, such an omission is less of an option. Now, treatment with drugs is tightly coupled, by law, medical practice, and insurance reimbursement schemes to DSM diagnostic categories. Problems regarded as "treatable" or as "symptoms" cannot float so free of the categories. So, while the lay media serve the earlier popularization function—identifying everyday suffering as treatable—they do so now by offering interpretations of disorder categories under which these struggles, now understood as loose congeries of "symptoms," are subsumed. In this way, articles after 1990 not only reflect the neurobiological imaginary, they serve to transpose a wider range of struggles into the arena of disorders. When people like Georgia identify with symptoms they found presented in mass media sources, this is the sort of transposition they appear to be referencing. When they reflect on the role of direct-to-consumer advertising, it is more explicit.

### Direct-to-Consumer Advertising

About half of those interviewed commented on direct-to-consumer advertising (DTCA), either volunteering information or answering a question when, given other comments they made about sources, it seemed appropriate to ask. Some of their reflections were fairly general, only loosely connected to their own experience. A few emphasized some general educational benefits of DTCA, for instance, such as that it promotes an understanding of conditions and which medications are available. A few were critical, suggesting that ads blur the boundaries between common experiences and mental illness. Others believed that these ads make some false promises. Margaret, for instance, complained that they "portray just this beautiful, wonderful person whose life has been completely fixed by this drug." And a few mentioned that the ads do not provide adequate information about side effects.[23]

Many people, however, engaged the drug ads on a more personal level, as giving them information about themselves. Some, including Georgia, directly identified with the symptoms presented in an ad—she "sees" what she is feeling in the depression checklist she encountered in advertising; her experience is "right there"—and this, sometimes along with other experiences, led to appropriating a disorder category. Gretchen, for instance, saw a DTC ad as encouraging her to take action. Age thirty-five and working two part-time jobs, she reported experiencing a lot of discomfort in social situations with people she does not know well. She had been to psychotherapy earlier and also said she had used the internet to look up support groups (though she never joined one or shared personal information), medications, and related information. Her "anxiety," too, she said, was "self-diagnosed," though she also notes the input of her boyfriend. When talking about the first time she went on medication, she specifically referenced DTC ads she had seen. And though she mocks the images, she acknowledges that they "get you thinking, maybe I should do something. . . . So, I went in and I asked to be on something."

Another type of information that people attributed to drug ads was the medicalizing explanation of chemical imbalance. Jon, for example, a senior in college, had experienced over the previous two years a set of problems that he saw as connected: not fitting in at the college he had transferred to, attention problems at work, and then a lack of interest and enjoyment in social activities. At first, he "self-diagnosed" (his term) with ADD and went to a psychiatrist, who put him on a stimulant. He also went to a psychologist, who thought he was suffering from depression, not ADD. He eventually came to accept the depression diagnosis and to adopt a highly medicalizing definition of his suffering. While he could not give any specific examples, he learned of the chemical imbalance explanation from drug ads, which, he thought, "say it a lot."[24]

In order to put these types of comments into context, it helps to again compare current practices with those of the tranquilizer era. In both time periods, pharmaceutical advertising has played an oversized role in how drugs, problems, and patients are represented. The new prescription tranquilizers of the 1950s were massively marketed, and as early as 1958, a committee of the US Congress was raising questions about the "false and misleading advertising" of these drugs. The Kefauver-Harris Amendments of 1962 permitted drug advertising only for FDA-approved uses—those disorders for which the drug had proven safe and effective—and required accurate information about side effects. As we saw with the FDA censure of the 1970 Serentil ad "for the anxiety that comes from not fitting in," the new rules limited advertising for prescription drugs to their use for "certain disease states" and proscribed pitches to physicians aimed at such off-label problems as "relatively minor or everyday anxiety."[25] Yet into the 1970s, ads could directly represent problems with irritability, loneliness, fatigue, isolation, a sense of inadequacy, disappointment, relationship conflicts, feeling overwhelmed, and so on to doctors as the sort of problems that medications could treat. But after the DSM-III, behaviors and emotions cannot be presented in themselves, so to speak, but only as symptoms of a DSM disorder.

In the earlier period, pharmaceutical promotion primarily targeted physicians. Companies spent the bulk of their marketing budget plugging drugs to doctors, through office visits, seminars, free samples, promotional giveaways, and advertising in medical journals. Although they certainly made efforts to influence popular media and generate public attention, they did not market their medications directly to consumers. That began to change in 1981, when the first DTC ad for a prescription drug appeared in *Reader's Digest*. Other such ads followed, prompting the FDA, which was troubled by the idea of DTCA, to ask for and obtain a moratorium in 1983. In 1985, while still holding the view that "direct to the public prescription advertising was not in the public interest," the regulatory agency allowed the ads to resume as long as balance and accuracy standards were met in the description of drug risks and benefits.[26] Then, in 1997, the FDA, adopting a more positive view of DTCA, issued new rules that significantly changed the requirements for disclosing side effects in broadcast advertisements. The new guidelines required only a "major statement" of the most important drug risks along with "adequate provision" for consumers to access the full package labeling. This liberalization made it possible for drug companies to air thirty- and sixty-second radio and TV commercials that had been effectively proscribed under the old rules.

It remains the case that the lion's share of industry marketing is to physicians. However, after the 1997 rule change, DTCA expenditures, which had been growing steadily in the 1990s, exploded.[27] Within a few years, nearly everyone had seen DTCA, whether in print publications, television, radio, direct mailings, and the like. More recently, these avenues have plateaued, and companies have increasingly emphasized so-called eDTCA: the internet and social media, such as Facebook, Twitter, and YouTube.[28] Early on, national surveys of both people who had recently seen a physician (1999, 2002) and physicians themselves (2002) demonstrated "a nearly universal awareness of DTC advertising."[29] These surveys showed, as have others since, that DTCA affects "consumer" behavior by prompting many people to pursue further information "about the drug, the condition it treats, or health in general."[30] In fact, as DTCA expenditures rose, patient spending on prescription drugs grew dramatically.[31]

DTCA, as with popular media and public educational efforts, has evolved over time.[32] The early drug ads for categories like social phobia, adult ADHD, and depression were very text heavy.[33] The condition itself, unfamiliar to the public, had to be explained. Social phobia (officially renamed "social anxiety disorder" in the DSM-5) was a new disorder with the DSM-III, little discussed even among professionals.[34] In the general public, there was, as the marketers say, little "disease awareness" of social phobia. Thus in 1999, when GlaxoSmithKline sought to market the SSRI antidepressant Paxil for social anxiety disorder, it began with "illness-awareness" or "help-seeking" advertisements.[35] This type of ad describes some general symptoms of a disease or condition and encourages viewers to see their doctor to discuss treatment options, but it does not actually mention a branded drug. The first campaign included a widely distributed poster with the slogan "Imagine Being Allergic to People." It featured a picture of a man whose appearance bespeaks dejection; he has wanted to join the conversation that is taking place at his table but could not. Text in the lower half of the poster defined the condition, from which "over 10 million Americans suffer," as "an excessive, persistent, disabling fear of embarrassment or humiliation in social, work, or performance situations" and offered a description of "what social anxiety disorder feels like." The "good news," according to the poster, "is that this disorder is treatable," and the ad offered an 800 number, a website, and the advice to "talk to your doctor."[36]

The same "awareness" process was followed in 2003 in the case of adult ADHD. ADHD, as defined in the DSM-III and in earlier concepts such as "minimal brain dysfunction" and "hyperkinetic reaction of childhood" in the DSM-II, was a childhood disorder whose symptoms would disappear in adolescence. Professionals had noted the phenomenon of "adult hyperactives"—those whose disorder persisted beyond childhood—beginning in the late 1970s, but the notion of "ADHD adults" was slow to catch on.[37] Only in the 1990s did adult ADHD begin to receive attention in the popular media, and only in the fourth edition of the DSM, published in 1994, is a place clearly established for this possibility in the diagnostic criteria. The first drug to win approval from the FDA for the treatment of adult ADHD was Strattera (in the ads, the condition is referred to as ADD). Prior to the full product-claim ads for the drug, Eli Lilly, the manufacturer, ran illness-awareness commercials with the theme "take control." They featured a woman, Anne, at an important staff meeting. While the boss is reporting that "sales are down" and "we've got our work cut out for us," Anne is shown with a blank, faraway look on her face. What is taking place inside her mind is represented by a series of scene changes that flash across the screen in rapid sequence, like the changing of TV channels. Some are innocuous (a flower, a bird, a traffic signal), and some depict Anne in distress (angrily tearing up a paper, reacting in frustration as she spills the contents of her purse). The voiceover explains that "adult ADD could be why it is so hard to sit still or remember appointments or obligations or why you feel distracted, disorganized, unable to finish things." The ad ends with the boss calling on Anne, who looks up startled and speechless while the voiceover says "ADD . . . a condition your doctor can help treat." The viewer is directed to a website to "take the first step": a short diagnostic questionnaire that you can print out and take to the doctor and thereby "take control."

While the early DTC ads for depression are full product-claim ads, they also give considerable space to explaining depression and listing some symptoms. The pharmaceutical firm Eli Lilly, for instance, ran ads in the late 1990s for the antidepressant Prozac with the theme "depression isolates." These ads, also text heavy, explain that "when you're clinically depressed . . . you may have trouble sleeping. Feel unusually sad or irritable. Find it hard to concentrate. Lose your appetite. Lack energy. Or have trouble feeling pleasure."[38]

Illness awareness is a kind of spadework, often requiring a lot of description and symptom lists. As noted, many patient groups, professional organizations, and federal agencies also engage in this practice. As with the DTC advertisements, the crucial component is to foster help-seeking by coaching people, through loosely worded symptom checklists and visual cues, to see a problem that they did not recognize before or to define a problem or experience in new, medical terms. Introducing the DSM disorder labels—or more accurately, user-friendly, less intimidating versions of those labels—provides the necessary language.

The DSM system implicitly presupposes that sufferers have such information available to them. Doctors alone cannot provide it: in fact, unless triggered in a doctor visit for another reason, some prior judgment of a problem is required to initiate medical help-seeking. In some cases, of course, such as with major mental illness, the unusualness and inexplicableness of behavior and emotion—delusions, hearing voices, suicide attempts, manic losses of control—may prompt help-seeking (often urged or initiated by others). In these cases, "awareness" does not need to be raised, and in fact few illness-awareness efforts or DTC ads are directed to such serious conditions.[39] However, consider the examples just discussed. How would ordinary people know that the experiences mentioned in the ads above, like "fear of embarrassment," "distracted," "disorganized," "trouble feeling pleasure," feeling "sad," or "vague aches and pains," are signs of a diagnosable psychiatric condition? The ads instruct readers/viewers to talk to a doctor precisely because otherwise it might never occur to them to do so.

As the conditions became better known, subsequent DTC ads for these conditions have challenged readers/viewers to see their ambiguous experience as a "real medical condition" (even the listing of bodily side effects adds to this "realness") that can be treated with the branded medication.[40] More of the message, like print and broadcast ads in general, is carried through visual images, and pictures of the promised new life are presented. These images generate meaning in the way that they and the text/voiceovers are organized and associated, both within the ad or commercial and with reference to wider social and cultural knowledge.[41] Sufferer stories, as visually displayed in the ads, are the representational core. Disorder names and symptom lists are still prominent, but the aim is even more explicitly to foster an identification through comparison, to offer reprieve, and to raise hopes, rather than to convey information.

An advertising series for the SNRI antidepressant medication Pristiq running in 2015 and 2016, for instance, had the headline question, "Does depression hold you back from enjoying life?"[42] A before-treatment drawing depicts a female windup doll, dressed in a drab-colored shirt and mid-calf skirt, slippers on her feet. She has a grim set of the mouth and a hunched posture and straight arms; her spring has wound down. She is looking longingly at a framed mirror, which does not reflect back her current condition but an after-treatment image. In the mirror, she is depicted standing on the platform of a carousel, dressed in sporty clothes, her arm extended to her smiling young daughter, who gazes up at her lovingly from the back of a bright-pink pony. Her husband is on the other side of the daughter, with one hand on her shoulder, the other on the carousel pole. His look is one of expectant welcome. The mother, now smiling, has just joined the activity in progress; she has been away and has been missed. The drug treatment was the turning point.

In print ads like this, the sufferer is shown outside of any particular context, an isolated and troubled figure. In others, and even more in television commercials, the images often seek to convey both a look and, perhaps especially, a feel of distress. Each is a sign system out of everyday life, with the "before" images showing or conveying familiar tropes—anxiousness before a meeting, sitting around the house in a bathrobe, looking on while others enjoy a party—and the "after" images showing pictures of success. The implication is that the condition has interfered with otherwise attainable social opportunities, work or school excellence, harmonious family life, romantic possibilities, and the like. The images, of generic and middle-class people in familiar settings, do not show any discernable difference between disorder and the vicissitudes of life, no sense that the action or emotion portrayed is particularly "excessive" or "marked," "persistent" or "irrational." Rather, these are just the sort of common experiences—of aloneness and sadness, feelings of frustration and embarrassment, a sense of being beleaguered and overwhelmed—that the old drug ads from the 1950s and 1960s showed, but now portrayed within a rubric of symptoms and disorder.

The stories of such common distress and limitation invite a transposition. These experiences are a part of our world and exhibiting them calls our status into question, invites our uncertainty, and asks us to attend to the possibility that we are missing something crucial about ourselves.[43] The images and the syntax aim to provoke an emotional connection with stories that overlap with our own, to bring to mind our own painful feelings of isolation or ineffectiveness or embarrassment. Even the "after" scenes—scenes of joyful reunion, new friends, work success, confident assertiveness, and the like—invite a comparison. The story these objects of desire tell is not so much about recovery from an illness as about social approval and living a more active, complete, whole, and successful life.[44] They signify the transformative power of the drug, but they also implicitly challenge the viewer: is your life all that you want it to be? All it *should* be? Perhaps, they intimate, you have made self-limiting decisions without understanding why. Perhaps you do not have the kind of relationship you *ought* to have with yourself.

The ads, of course, never actually say you have a particular condition. Rather, they place you in a liminal state: your experience "could be" an indication of the condition. You could be calling your experience by the wrong name; you could be missing, as the Strattera commercial for the treatment of adult ADD put it, "the explanation you have been looking for all your life." You could be hurting yourself and others. And this possibility creates an imperative to take immediate action: "Take the next step"; "Call your doctor today."

Further, and importantly, "a chemical imbalance *could be* to blame." The dynamic of motivating viewer appropriation of a diagnostic category involves more than an effort to create or heighten unease. Once provoked, the ads also work to relieve some of that tension and provide a sense of reassurance and meaning. One form of relief comes in the common references to a chemical imbalance. In the context of DTCA, a neurochemical imbalance is good news. The distress, as listed in symptoms and portrayed in the sufferer stories, is always presented without context. It has no story; it is simply there, ominous and injurious, waiting to be treated. The chemical imbalance explanation establishes the condition as a "real"—organic, biological—"medical condition," yet also different from mental illness. There are no references in DTC ads to psychiatrists or the DSM, no use of phrases like "mental illness," or "mental disorder." Rather, conditions are framed as the sorts of somatic difficulties regular physicians treat, their ordinariness distinguishing them from psychiatric disorder, complicated circumstantial dilemmas, and the psychologically enigmatic.[45] The meaning of experience is straightforward, right on the surface.

Other possibilities, explanations that might reflect negatively on the self and its autonomy, are ruled out. The sufferer is *not* to blame because the imbalance is an issue solely of the brain—effort, psychological issues, evaluative stance, self-understanding, morality have no part. Despite the seeming introduction of a biological determinism, the ads address the reader/viewer as a free agent, a self-determining subject of choice and optimization. Neurochemical imbalances are also presented as quite common. DTCA and educational campaigns reassure sufferers that they are not alone, emphasizing that disorders like depression, attention deficit, and social anxiety are very widespread, affecting millions. And they are medically treatable. In DTCA, the prognosis is positive because there is nothing mysterious going on that the medication cannot correct. The pill alone is efficacious. Other interventions, such as psychotherapy, are almost never mentioned.

Like the medical ads from the minor tranquilizer era, what DTCA presents are portraits of distress, opportunity, and the power of medications. Painful experiences are what defines the "disorder" referenced in the ad, not some abstract DSM criteria or definition. They are the signifiers that the ads invite the readers/viewers to exchange for their own experience. Some interview participants spoke of print ads and commercials that they "related to," "identified with," "resonated with," and the like. In identifying their own experience in what they saw represented, they make the medical connection. This exchange is not simply a perception of what was written or seen in ads. It is a product of their mind, their imaginative participation.[46] Like we saw with Georgia, it becomes *their* name for their embodied experience. They imbue it with personal meaning and integrate it with their self-understanding. The medication follows.

The portraits of distress and new life in DTCA are not just for sufferers. They are also for doctors and other health professionals.[47] And they are clearly directed to people who might identify the portrayed distress and limitations in someone close to them. Most of the time, this is implicit, but it is occasionally expressed directly. In the above-described illness-awareness commercial for adult ADD featuring "Anne," for instance, a secondary audience is indicated: "If you, or someone you care about, has felt these symptoms . . ." The marketers know full well, as I'll discuss next, that family and friends commonly play an important role in encouraging help-seeking and treatment.

## Family and Friends

For most people, making sense of and coping with the challenges and predicaments they face in their daily lives is not a solitary activity. With only two exceptions, participants recounted talking with at least one other trusted person close to them—a confidant—whether a family member, spouse, friend, or acquaintance, about the struggles they were facing.[48] These confidants influenced their practical decisions and interpretations. Kristin, from the introduction, reported meeting a young man at a party who talked openly with her about his diagnosis of ADHD and success with medication. That encounter persuaded her that they "had the same sort of thing" and "fit the same mold." Eric, from chapter 2, identified his wife as confirming that his emotional responses were disproportionate and encouraging him to seek medical attention. Georgia, discussed above, spoke of a close friend to whom she compared her stressful experience and whose encouragement to seek medical help and medication helped convince her to follow through. Jon, also mentioned above, noted that his understanding of depression and "SSRIs," as he referred to them, was influenced by his mother, who originally supplied him with antidepressant pills she had on hand.

These examples illustrate patterns that appeared in the stories of many people, who reported being encouraged, but not pressured, by confidants to seek help or treatment, whether they followed the advice or not. Of those who had seen a professional (two-thirds of all participants), just over half reported that a friend or family member had backed their initial decision to seek help.[49] The remaining cases involved other patterns, including some where a professional, such as a doctor, encouraged the help-seeking and a few that involved students who went on their own initiative to college health centers, a free service whose use colleges actively promote. Even among those who had never sought professional help, about a quarter reported that a friend or family member (eight people) or a doctor (one person) had suggested to them that they might benefit from it. In all but one of these instances (which was unclear), the help that was suggested—the help they did *not* pursue—was psychotherapy.

In many of their stories, participants described their confidant's influence on help-seeking as mainly confirming a problem they already knew about—a "wake-up call" as one participant put it—and so helping them overcome reluctance to contact a professional. In some cases, the interaction was also critical to facilitating appropriation of a diagnostic category or generating a determination to seek medication (only two identified confidants as directly dissuading appropriation). Confidants helped with interpreting troublesome emotions or behaviors as unusual or creating limitations that need not be endured. In another role, confidants served as direct models (generally, the comparison models are implicit) with whose experience sufferers compared and evaluated their own. In some cases, these models were parents or siblings.

A particularly stark example is Becca, a senior in college, who was first diagnosed with depression and put on an antidepressant at the age of thirteen. She recounted learning about depression from her grandmother, who was "very mentally ill," when she was young. The grandmother told Becca that both she and Becca's mother had depression and so Becca might get it too, but help was available. Years later, when Becca was in the eighth grade, she "began feeling incredibly down and hopeless." Although she said that "there really wasn't a reason for it," she also said that middle school was a "social hell" because she was bookish and shy and felt out of place. Whatever the circumstances, she was talking with her mother one day and told her that life had been hard for several months and that she thought she had depression. In response, her mother provided her with an antidepressant tablet. The very next day Becca felt great, and so her mother took her to the family doctor, who diagnosed depression and put her on medication, which Becca was still taking at the time of the interview nearly ten years later (and over the concerns of her doctor who has suggested she stop).

There was more talk of "we" in Becca's story than in the other cases where the participant first saw a professional as a minor. In the others, which all involved a diagnosis and a prescription, the mother is generally presented as the initiator and the adolescent as somewhat passive or ambivalent. However, in each case, the young person came to appropriate the category, and in all but one, in which the prescription was never filled, all remained on medication. Parents, spouses, or siblings were also important for a number of participants whose diagnosis came as an adult.

As with Becca, confidants may not only alert participants to a diagnostic category but to the promises of medication. In fact, in every instance but one—which led to psychotherapy—this was the pattern. Linda, for example, fifty years old, saw her problems with feeling down and overwhelmed as part of an "emotional midlife crisis." Although she had always worked part-time and was never the "typical suburban mother," the crisis began, she said, with the anticipation of an "empty nest" when her youngest went off to college. She was also feeling the strain of a deteriorating relationship with her husband. She first began thinking about taking medication when a friend mentioned the possibility:

> I guess I didn't realize that I was in a downward spiral, and then . . . I was having dinner with a friend one night, and I wouldn't even consider her a real close friend, but we were walking out of the restaurant and I must have said something again, and she said, "you know you can take medication for that, don't you?" And . . . for some reason it just hit me on the side of the head—you know, I don't have to feel so . . . overwhelmed or sunk. So, I went to my sister's doctor, who is a psychiatric doctor I guess, and she gave me a prescription.

The psychiatrist wrote a prescription but also insisted that Linda go to counseling, which she did for a year or so until life got too busy. She has remained on a medication.

Again, as with medical professionals, I have no direct knowledge of what participants' confidants might have actually said or done. The role patterns described, however, were fairly consistent across groups of participants and not at all idiosyncratic. General studies of help-seeking also demonstrate the important role that intimates play in encouraging people to label and address a problem.[50] The very fact that nearly every participant who saw a medical professional mentioned at least one trusted confidant is significant. It certainly speaks to the basically dialogical nature of the appropriation process. Symbolic representations in media and advertising were important, but in only one case were they enough in themselves. And medical professionals, for the most part, were entrusted with—or allowed or took—a lesser role. Although the circle of confidants was very small, in many cases category appropriation and willingness to take medication appeared to need this enabling feedback.

The role of confidants was also significant for another reason. Throughout the interviews, people seemed to be expressing a felt need for some "witness," some external confirmation of the rightness of their subjective views and treatment decisions. I introduced this idea earlier with respect to doctors. For those who appropriated a category, the doctor's diagnosis was prized for conferring an objectivity and legitimacy on their struggles and medication-taking. In a more general way, participants also appeared to both seek and need the validation of at least one other person. Beyond the roles already discussed, they also referenced family and friends at other crucial junctures in their stories, sometimes invoking them as support for their views, evidence for the reality of problems or the absence of disorder, justification of the need for help, and confirmation that their medication was working. They were making decisions in a discursive field that included alternative, nonmedical interpretations and interventions, and in these instances, confidants were involved in the negotiation of doubts and resistance. Witnesses were, in a sense, part of the participant's "team."[51] At the same time, and with respect to many of the same issues, others who might not be counted upon to provide support in the desired way were left out of the dialogue.

## Conclusion

Those who appropriated a disorder category—and with it, in most cases, a medicalized explanation and medication—found an impression point in interaction with several key sources: medical professionals, popular media and the internet, DTCA, and friends and family. It will now be clear that these sources—including, perhaps, the professionals—operate at a semantic distance from strict psychiatric concepts and ways of knowing. That does not mean that the messages are always at odds, but simply that popular information is coming in real-world and nonscientific registers, with different practical purposes. When people encountered ostensibly psychiatric language, in other words, they were encountering it not in some abstract, experience-distant, or neutral dictionary-like sense, but as used in specific, problem-solving contexts.

Put differently, participants finding the impression point that transformed understanding was not a matter of passively accepting a formal diagnosis, a disorder identified by a doctor and presented to them. A formal diagnosis was important, of course, for both symbolic reasons and for access to medication.[52] But appropriation and formal diagnosis were different things and could be completely separated, as when people appropriated a category without ever receiving a formal diagnosis. The wide popularization of medical concepts like depression, social anxiety, and attention deficit have made them virtual "floating signifiers" on the healthscape, no longer simply the province of medicine or requiring its authority.

Participant impression points were fostered by a whole series of assurances and promises aimed to reduce resistance, generate commitment, and instill hope. Doctors, media, and ads are not the same, but they converge around a common message:

- Your everyday struggles are symptoms of "real medical conditions."
- You are not alone; these conditions are widespread, and virtually everyone is susceptible.
- You are not to blame; these conditions are caused by neurochemical aberrations that are outside conscious control.
- You can be (easily) treated; drugs work to correct the underlying somatic malfunction.

And, more implicitly, at least in many media sources and DTC ads:

- Your inner life and evaluative stance are of marginal, if any, relevance; counseling or psychotherapy aimed at self-insight would serve little purpose.

Confidants often convey the same information and feedback, some serving as models and witnesses to a biomedical orientation and reinforcing the rightness of seeking expert help and taking medication. And across the board, what is being offered is not some cure for an illness but the promise of a life in which one's status as a certain type of person is restored or enhanced.

That promise is the key. Over the past two chapters, I have explored aspects of the grammar of the neurobiological imaginary—DSM categories, the new biological language, and ideas about treatment and how these are popularly conveyed to ordinary people. When people speak of their everyday suffering, its causes, and the means to ease it, they often reference, both directly and implicitly, this basic grammar, even when rejecting it. The biomedical healthscape has become all but unavoidable. Yet, the availability of this grammar does *not* determine how it is taken up or to what ends. And it carries its own challenges.

In a predicament, a disturbed selfhood is at stake. Revealing one's suffering to others (including professionals), appropriating a diagnosis, taking a medication, interpreting experience in terms of biology—all have public meanings and potentially negative implications for self. Participants did not take these lightly. They devoted considerable effort to contain risks and interpretatively neutralize unfavorable implications while they forged meanings in terms congenial to their desired self-image, needs, and purposes.[53] These meanings reflect prominent cultural norms and ideals of a good self, a particular ontology of being. What gives the neurobiological imaginary its force is not the "biological" as such, which, after all, is little more than a language of disengagement. The force is in the synergy between the interpretation opened up by the imaginary and the normative picture of what it is good to be. Imagining suffering in neurobiological terms gives this picture a sense of realizability. I begin considering this appeal in chapter 4 by looking at how people resist implications of differentness.

# FOUR

# Resisting Differentness

In her senior year of college, Jenna discovered that her boyfriend was cheating on her. They had only been dating for a short time, yet, upon discovering the cheating, Jenna went into an emotional tailspin. She was "crying all the time," she says, lacked motivation to go to classes, was sleeping a lot, and avoiding situations that she normally enjoyed.

Jenna attended a selective university, where she had been a good student and a dedicated athlete. Several injuries brought her athletic career to a premature end after her junior year. Because running had been a big part of her "former personality," she had a really hard time dealing with that change. Then, six months later, came the betrayal and sudden breakup with her boyfriend.

In describing her emotional reaction, Jenna stresses its unreasonableness. Given that they hadn't dated for very long, she felt that she shouldn't be so upset and that a month—the point at which she went to see a counselor—was an inappropriately long time to be sad and still crying about the matter. Even more disturbing was her self-blame and sense of personal failure. Jenna had invested emotionally without a commensurate commitment by her boyfriend and now found herself focusing not on his fault but on her own actions. "Instead of being, 'How could he have done that to me?' and 'I deserve better,'" she says in describing her immediate reaction to the cheating, "It was, 'What did I do wrong?' and 'What can I do better?'"

Jenna's lack of confidence and inability to quickly move beyond the relationship was a sign to her that she was simply not herself. In fact, she knows how she would have reacted had she been in her normal, rightful state of mind. She explains:

In the past, even though that [infidelity] had never happened to me, that's
not an approach I would have taken. . . . I would have had self-confidence
to be like, "Well, that's his loss and I'm better than that," instead of taking
all the blame on myself even though he's the one that messed up. So, I think
I had a hard time seeing that. My friends kept telling me that and [it] just
wasn't, like getting in there.

Something was seriously amiss, and she decided she needed a "third-party"
perspective.

Jenna first consulted a close friend about going to counseling. The
friend strongly encouraged her to do so and shared that her own parents
regularly went to family counseling. Jenna felt supported but still found it
difficult to make the appointment. It meant admitting that she was having
problems, and she worried about the reaction she would get. She said,

they're going to think I'm crazy or not crazy yet. [laughs] I don't know how
to explain that. It was like, I don't want them to think that I'm nuts but at
the same time, what if they think that I'm just not sucking it up and dealing
with life and I'm being too weak about it or something. So, I sort of had two
different ways that this could go.

But after confiding in her parents, who added their encouragement, she
went to the student health center. There she found talking with the counselor much easier than she had anticipated.

Jenna had already seen an antidepressant advertisement that showed,
she says, "exactly how I feel," and she and the counselor discussed medication at the first appointment. They decided to begin with some talk therapy, but after a few weeks, the counselor referred her to the in-house psychiatrist. Jenna was getting into conflicts with her friends and parents, and
while the psychotherapy was helping, she remarks, "we needed to take the
next step to drugs." The consultation lasted about two hours during which
Jenna filled out questionnaires about her experience. The psychiatrist then
quickly made a diagnosis of depression and anxiety and prescribed both
an antidepressant and a sleeping pill.

Having been confused and disoriented by her emotional responses,
Jenna found receiving the formal diagnosis of depression (she makes no
reference to the anxiety) a relief. It provided an explanation that helped
her to acknowledge that she was "having issues." It confirmed that there
was really something wrong and that she was not, she says, "just a slacker"
or the sort of person who doesn't try hard enough or do what is expected

of her. At the same time, the diagnosis didn't seem particularly threatening since depression is so common and her own case was on the milder end of the "illness continuum."

Along with the diagnosis and prescription, the psychiatrist offered a biological account of her distress. He spoke about serotonin levels and showed her some brain scan pictures with bright colors that helped convince her, she recounts, of the "whole imbalance of brain chemicals" explanation. Though she doesn't recall any of the specifics of what the psychiatrist said, his account helped her to understand, "from a scientific standpoint," why she "was just crying over nothing." It confirmed an impression point.

After graduating, Jenna sought out another counselor, though the sessions quickly tapered off in frequency and focused primarily on medication management. She wanted some counseling because she was worried about undesirable side effects (which she has experienced with the sleeping pills) and because the effects of the antidepressant are subtle and hard to read. "It's not like when you have a cold or something," she observes, "and okay, the cold's gone, like it's obvious to tell." She needed an informed, outside opinion to help her interpret changes and ongoing progress, something she regarded the psychiatry appointments as too limited to provide.

Subtle or not, Jenna, now twenty-three, believes that the medication has been essential to her improvement. Though she feels her creativity is dulled, she is pleased that her emotional reactions are less extreme and her moods more stable. She sleeps better and has more energy and emotional control. She is, she says, "very close to functioning the way I used to." She expresses a strong need to continue the medication in order to feel normal and worries that without it she could "get worse and worse to the point where I . . . would cease to exist and be myself and I would be completely lost." Despite this worry, she wants to have a long-term plan to stop.

---

As Jenna's story illustrates, a predicament can involve a series of critical junctures at which decisions are made that involve other people. Each of these decisions carries both risk and promise—practical, explanatory, and existential. As she began to spiral downward, there arose the question of with whom to speak or seek advice. Jenna began with her best friend and then her parents, whom she has always listened to and trusted. They confirmed that her reactions and self-blame were anomalous and that seeking help was the right step. At least initially, Jenna responded to her other roommates, who noticed her changed behavior, with "excuses." But after beginning to see the counselor and get some handle on her experience, she extended her circle of confidants to include these friends and found some comfort in their understanding. They "were actually really good about it," she notes, "because they would give me positive reinforcement without being over the top"—that is, without making too much of her improvement and thus of the behavior that preceded it.

For Jenna, the question of seeking help and of talking with her parents was made somewhat urgent by the fact that her classwork was suffering and she was nearing graduation. But the decision was still a challenge, and the affirmation of her confidants was important. Pursuing counseling also raised the question of how she would be regarded by the professionals. They might make evaluations contrary to her own. She could be perceived as "nuts," or she could be perceived as "weak." Disclosing her emotional problems made her vulnerable and carried the distinct risk of drawing undesirable interpretations.

Then there was the decision about treatment. Getting the third-party perspective of the counselor turned out, seemingly, not to entail any unfavorable evaluation or interpretation alternative to her own. She says the talk was helping her, though she does not say how, nor mention any psychological insight, struggle, or change of view. But though she was already open and even predisposed to medication, turning to drugs was, as she says, "a much bigger step" than therapy. For Jenna, medication had larger implications for her self-control, and pursuing it introduced medical issues of diagnosis, being "sick," and physical effects, both positive and negative ("side effects").

For Jenna, her emotional reaction and self-doubt are jarring, out of place: they defy her basic assumptions about herself as confident, mature, and self-sufficient, about who she is and the trajectory of her life. The formal diagnosis, linked with the notion of a chemical imbalance, provides an explanation that makes sense to her and one she finds reassuring. The problem could have been something harder to reconcile: "I would have been so much more disappointed," she says, "if it's just me being crazy in my head and not being able to function or something." The neurobiological disorder confirms that the problem is largely outside of herself, that she would have conformed to expectations had not something gone chemically awry. She now sees the prognosis as positive; her experience can now be explained in a cause-effect sequence and the imbalance treated. The diagnosis also helps her explain herself to others and brings some "sick role" benefits of adjusted expectations from her parents, roommates, and teachers.[1] And the diagnosis has brought access to the medication, which she credits (with help from her counselor) for her increased happiness and functioning. It is helping restore her to her "former self."

At the same time, Jenna sees mental illness as implying a loss of control, and during the interview, she is concerned to distance herself from that implication. She comes to speak of herself as "sick," for instance, but emphasizes that she is not at the end of the continuum with the "really sick people." She appropriates depression in terms of a chemical imbalance, but stresses that it is not that serious. And with respect to medication use, she maintains that she had no real choice. The therapy just wasn't enough, and she offers the witness of the counselor as confirming this judgment. Without the help of the drug, she would have gotten "worse and worse." Further, she is a very responsible, proactive patient: she is compliant with the doctor, has engaged a third-party counselor to assist her with evaluating drug effects, and plans to get off the drug at the appropriate time. In short, she is not giving up or taking the easy way out. She is working hard and positively engaging her necessary treatment with energy and initiative. She is, she intimates, already acting like her former self.

Jenna, like others we have seen, is acutely aware that aspects of help-seeking, diagnosis, a medicalizing explanation, and medication treatment can spoil self-image and social status. Some of these dangers people spoke of directly. They named specific threats that they were aware of and concerned about, emphasizing their agency in negotiating those threats and careful to control who learned about them and on what terms. Other threats lay more in the background, not directly expressed but clear from their stories and the standards and obligations against which they compared themselves and sought to conform. But in either case, they sought to neutralize or contain these dangerous elements.

This effort is not a negative movement. Everywhere I turned in writing this book, from popular sources to drug advertising to what participants recounted from their doctor visits, I encountered the idea that what motivates the turn toward the medicalizing perspective is blame—the desire not to attribute blame to self or others. This quote from a journalist nicely captures a widespread view: "The notion that depression is a biological ailment, like Alzheimer's, proved enormously appealing to patients. It relieved much of the stigma of mental illness, which could now be viewed not as a personal or moral failing, but as a glitch of biology."[2] In this view, the danger surrounding mental health problems is almost exclusively a matter of negative stereotypes or "deeply discrediting" attributions imposed by society on (and then perhaps internalized by) members of certain groups or categories.[3] Sufferers, one such category, are subject to many negative images, such as that their suffering reflects poor character or a moral lapse. It follows that if the problem is in the brain, not in the mind, then the sufferer cannot be held morally accountable. For "the body," observes anthropologist Tanya Luhrmann, "is always morally innocent."[4]

True, some participants referenced guilt feelings and blame, and some expressed relief from these feelings in biological explanations. We'll have to consider what they meant. But blame, or a desire to "jettison the burden of responsibility,"[5] is not what makes imagining suffering and self in neurobiological terms compelling. Over this chapter and the next, I will argue that participants' meaning-making is an effort to impose an interpretation on their experience that both preserves and restores their autonomous selfhood. The goal is to address the basic questions—What went wrong? Why did this happen to me? How do I move forward?—so as to establish or reaffirm themselves as conforming to specific ideals and normative standards of self, emotion, and relationship. The neurobiological imaginary is bound up with how this good self is pictured and might be achieved.

But there are dangerous elements in medicalized explanation and practice. These are not principally a matter of social stigma; they are further threats to autonomous selfhood. This selfhood is a normative way of being that is breached by the anomalous experience of everyday suffering *and* by certain medicalized efforts to achieve or restore conformity.[6] The latter carry implications of disorder, differentness, dependency, and determinism, which is why adopting a medicalizing perspective was a fraught process for many and an important reason why others rejected it. The force of the neurobiological imaginary is in its synergy with self-defining selfhood. But only, as I'll detail in this chapter, with the medicalizing explanation and medication shorn of their unfavorable meanings. Those too had to be self-defined.

Because the emphasis on stigma is so common and seemingly intuitive, I begin with a discussion of the frequent and ongoing public campaigns to address it. These campaigns hold, to quote the antistigma literature, that "stigmatizing attitudes" and "judgmental assumptions" about mental illness, based on "myths," are the crucial obstacle to people seeking and receiving medical help for mental health problems. If only the lay public could realize that such problems—including everyday suffering—are just another form of medical problem, then attitudes among both sufferers and nonsufferers alike would be transformed.[7] Yet, after decades of these campaigns, no such benevolent orientation has emerged. If anything, the new lay openness to biology and drugs has hardened attitudes.

Considering this counterproductive result will help to highlight the nature of the problem with medical language and medication that participants sought to manage. Judgmental assumptions about everyday suffering and mental illness are real enough, and people were certainly conscious of them. But, as in Jenna's case, attributions of "badness" and moral responsibility were not the shoals that participants were especially concerned to navigate. For most, a more existential understanding of self was in the balance, a question of "differentness" that touches on self-worth and social standing. Medical language and medication neither suspends nor alleviates this concern. Rather, each creates challenges of its own.

## "A Flaw in Chemistry, Not Character"

Public campaigns to reduce the stigma associated with mental illness have a long history. As early as 1951, the Department of Public Health in the Canadian province of Saskatchewan conducted a six-month intensive education campaign in a small town to try and change the "negative attitudes of the public toward the mentally ill."[8] The project team, led by a psychiatrist and a sociologist, worked with the local newspaper and various community leaders and organizations in an "all-out" effort to reach people through showing films, sponsoring school programs, distributing pamphlets, airing radio broadcasts, creating small group discussions, and more.[9] Their message concentrated on showing the fuzziness of the line between normal and abnormal and on fostering the belief that "behavior is caused and is therefore understandable and subject to change."[10] They sought to convince people to adopt a psychiatric perspective on mental illness and on human conduct generally. This meant teaching them, on the one hand, to stop "normalizing" a wide range of "disturbed behavior"—to recognize it instead as mental illness, not mere eccentricity.[11] Yet it also meant teaching them, on the other hand, not to "exaggerate" the difference between normal and abnormal. Not, that is, to draw a sharp, rigid boundary between those who did get labeled "disturbed" and everyone else. And, finally, it meant teaching them that the "mechanisms" of mental illness, just like those behind other kinds of illness, are understood and can be successfully addressed. If people could be convinced to adopt these "permissive" and "'correct' ideas," then they would turn from a "pattern of denial and isolation" and see that the mentally ill "were pretty much like everyone else" and need not be the subject of blame or met with fear and avoidance.

This early campaign was a comprehensive failure. Using questionnaires and interviews of townspeople, both before and after the campaign, the project team found "no appreciable changes in beliefs about mental illness or attitudes toward the mentally ill as a result of the program."[12] In fact, the primary result of the campaign was to generate hostility in the town to the project and project team. The residents were far more tolerant of unusual behavior than psychiatrists and thought of mental illness as limited to serious conditions, on the order of "psychosis."[13] They rejected the program's postulate that "there is little difference between illness and health," had their own views of the causes of mental illness, and were far less optimistic about recovery. They appeared to question the belief that the mentally ill are "really ill in the same way as the physically ill," which the team leaders identified as the bedrock of the "modern, rational, scientific approach."[14]

In the intervening years, there have been a great number of antistigma campaigns, with names like "Open the Doors," "What a Difference a Friend Makes," "Make It OK," "National Anti-Stigma Campaign," and "Time to Change." Recent social media efforts by sufferers to raise awareness have used hashtags like #imnotashamed and #sicknotweak.[15] The shared and salutary goal of these sorts of campaigns is to challenge stereotypes and change negative attitudes so as to build support and resource allocation for people with mental illness diagnoses, break down barriers to open communication and help-seeking, end discriminatory practices, and promote treatment.

The tenet that "mental illness is just like any other illness" has remained the bedrock principle.[16] The National Alliance for the Mentally Ill, for instance, which conducts regular "stigmabusters" campaigns, describes mental illnesses as "medical conditions that disrupt a person's thinking, feeling, mood, ability to relate to others and daily functioning. Just as diabetes is a disorder of the pancreas, mental illnesses are medical conditions that often result in a diminished capacity for coping with the ordinary demands of life."[17] In 2010, to give another example, the actress Glenn Close and Foundation House, which provides services to patients, launched a national antistigma campaign, called "Bring Change to Mind." The campaign produces educational resources, including public service announcements featuring celebrities. Their webpage states, "The fact is, a mental illness is a disorder of the brain—your body's most important organ."[18]

Emphasizing this somatic conceptualization also continues to be regarded as an integral step toward fostering a more "benevolent, supportive orientation" toward the mentally ill. But what, precisely, is the link between the two, between mechanistic model and benevolent attitude? The logic of the connection, I want to suggest, rests on an assumption—one might even say a wager—about the role of personal responsibility in mental health problems. The logic has three dimensions.

First, according to this logic, the root cause of the stigma attached to mental illness lies in misguided attributions of responsibility by both sufferers and nonsufferers alike. Both mistakenly believe that problems are "the result of moral failings or limited willpower" rather than "legitimate illnesses," to quote a 1999 US surgeon general's report on mental health.[19] Combating stigma must begin by undercutting this (allegedly common) moralizing explanation. Second, treating a mental health problem as a somatic disorder or disease will remove blame and responsibility. As one group of researchers express the idea: "public understanding of the biogenetic aspects of psychopathology will alleviate [mental illness] stigma by reducing the tendency to hold persons experiencing disorders responsible for their illness."[20] The responsibility will be reduced or eliminated because an explanation will be made with reference to some underlying dysfunctional mechanism in the individual that reduces or eliminates volitional control. As the "Bring Change to Mind" campaign states, "And just as with most diseases, mental illnesses are no one's fault. The unusual behaviors associated with some illnesses are symptoms of the disease—not the cause."[21] Third, once the burden of responsibility and the blame it engenders is lifted, sufferers will be open to seeking help and accepting and staying in treatment. This, too, will have the consequence of improving how other people react to them, over time creating a virtuous circle.[22]

In the Saskatchewan experiment, which worked from a version of this wager, the strategy failed. A follow-up study, done a quarter century later in the very same town, still found no change.[23] But antistigma advocates were undeterred and actively supported the biological turn in psychiatry. These advocates have celebrated the public's increasing endorsement of somatic causative factors, professional help-seeking, and "evidence-based treatments" (read, for the most part, medication, and then some CBT) for mental health problems. This endorsement means that laypeople are at long last becoming "literate," with a "more scientific" understanding of mental illness.[24] And now more knowledgeable in this sense, it follows that they will readily adopt the more "liberal, knowledgeable, benevolent, supportive orientation toward the mentally ill."[25]

Yet despite the growing public embrace of biology, the antistigma strategy continues to fail. Studies of the relationship between brain-disorder beliefs and stigmatizing attitudes do not find greater tolerance—not among psychiatric patients, the general public, or even among professionals themselves.[26] In fact, if anything, they show just the opposite. Granted, surveys find that increased endorsement of biogenetic beliefs correlates with greater acceptance of professional help-seeking and the use of prescription medicine, as well as some reduction in attributions of responsibility: three of the central goals of the medical and antistigma campaigns.[27] However, endorsement of such biogenetic beliefs also correlates with *more* pessimistic, stigmatizing, and social avoidance attitudes toward those with mental health problems, including by sufferers themselves.[28] Contrary to the theory, removing responsibility by recourse to a technical, mechanistic language results in a less favorable and more patronizing view of sufferers—and as a result, more stigma and isolation.

What accounts for this counterproductive result? According to interview participants, the message that "behavior is caused," to quote the campaign from 1951, is precisely the feature of a medicalizing explanation that is most difficult to accept. The mechanistic removal of responsibility and control calls their agency and social standing into question. In story after story, people distanced themselves from the loss of control imputed to serious mental illness. This was not a matter of lay illiteracy or social bias.[29] Reducing thoughts and emotions to the outworkings of physical phenomena intensifies the predicament. Michael, who experiences anxiety in social situations, explained why he hoped he does not have a mental disorder: "I guess having a disorder it would make you somehow not a bad person, but just a different person." He did not want to view himself as less than fully competent, as less than a normal member of the community.[30] One reason why the wager at the heart of the antistigma strategy is self-defeating is because the benign offer of medical absolution comes at the price of an implicit dehumanization.[31]

As Michael's comment suggests, "badness" was not actually a central issue. In the interviews, it was often unclear how much participants who spoke of fault actually blamed themselves for their predicament. In her initial reaction, Jenna, for instance, blamed herself for the breakup, wondered if she was making a fuss over nothing, and worried that she might be losing her grip. But early on she came to see her self-blame as part of the anomalous reactions that needed to be explained. No participant, in fact, saw the matter as a zero-sum dichotomy of moral failure versus a glitch in biology. The alternative explanation to biology, which was especially distinct in the stories of those who changed their mind, was not moral failure but psychosocial factors.[32] Some people, as we saw in chapter 1, found some fault with themselves for not responding better or doing enough to rise above their predicament. But in locating the cause of their suffering, they did not identify intentional action or inaction.

People perceived themselves to be in a predicament not of their choosing, in a situation where they had little immediate control over their ability to meet the normative standard or ideal. As Goffman observes, "mere desire to abide by the norm—mere good will—is not enough, for in many cases the individual has no immediate control over his level of sustaining the norm."[33] For participants, their suffering was not a matter of choice or intentional deviance, and thus they did not take the view that their predicament could be explained as simply a "moral failing" or, even more pointedly, something for which they could or should repent or seek forgiveness.[34]

The major concern for participants was not with being a "bad person" but with being a "different person," both in how others viewed them and, reciprocally, in their own understanding. While this unintentional differentness was not a matter of "fault" or "blame" in the moralizing sense, it did involve a sense of being tainted or deficient in some important sense relative to social norms clustered, as I'll elaborate in the next chapter, around the ideal of autonomy. Jenna's reaction to the breakup, for instance, was inconsistent with norms of self-definition, emotional self-sufficiency, and independence. It constituted a challenge to her self-understanding and sense of control that was deeply disturbing because the norms of autonomy imply the possession of a will that is sovereign and even invulnerable. As participants made clear in their own ways, evaluating their predicaments against norms of autonomy meant that their predicaments must in a crucial sense be self-inflicted; their trouble had to have involved some choice. Something must be wrong with them, with their *self*, imagined in this autonomous way. This was the most basic threat. Efforts to reconcile disparities, including those that arose in the process of help-seeking and treatment, were efforts to reaffirm or reestablish a self-conception and sense of autonomy that had been marred by trouble. When participants spoke, like Jenna, of reassurance or relief in a diagnosis or a medicalizing explanation, they were talking about easing the sting of this sort of existential challenge.[35]

The antistigma strategy is tethered to the medical model, which, as briefly noted in chapter 2, conceives of disorder and disease as discrete malfunctions in the individual body that arise from underlying pathological mechanisms. In the model, only the theoretical descriptions of professionals have truth value. They are based on scientifically discovered and value-neutral facts. The knowledge and descriptions that laypeople produce in their everyday practical activity, by contrast, are at best "deficient approximations of professional beliefs."[36] At worst, they involve moral or evaluative judgments and are "fit only for debunking."[37] In the medical model and the "disease like any other" idea, sickness is just a fact of the physical body: people get sick, they (should) get treatment, and then, if treatment is effective, they get (to varying degrees) better. That is all there is to the matter, whether the problem is with the pancreas, as in diabetes, or with thoughts, behaviors, and emotions. The effects of the medical condition, the symptoms, can of course cause discomfort, inconvenience, pain, and the like, but questions of "differentness"—questions of social norms and self-worth—are out of place.[38] The lay tendency to raise them is a category error.

Yet, raise them they do, whether they seek medical help or not. Even for the participants who referenced genetics and neurochemistry, more than some malfunctioning of the brain (body) was at stake. Talk of the brain did not render questions of self and social status superfluous, nor did it suppress participants' active interest in and efforts to manage the interpretation of their experience. Far from it. Each "moment" of decision making, as noted above, carried practical, explanatory, and existential risks and promises. In what follows, I'll consider the decision moments and the discriminations that people made about diagnostic categories, medication use, and disclosure, and I'll do so within the context of the disorientation and the difference that participants recounted in their stories. These moments are not reducible to a generic, negative struggle against social stereotypes; they are bound up with a deeper and positive effort to reestablish order and self-worth within participants' ongoing projects of self-definition. In the next chapter, I will carry the exploration of order further and consider the broader background norms of relationship, emotion, and self.

For most participants, their effort to find a voice began with whom they told.

## Disclosure

At the time of the interview, most participants were not in the initial throes of a crisis. While they still might have much that was unresolved, they had had time to reflect on their predicament, talk with persons close to them, and perhaps seek professional help. In telling their story, they spoke of one or more earlier crises, times of confusion when they were most intensely struggling to make sense of anomalous emotion and behavior, deciding to seek (or not to seek) medical help and treatment, or asking themselves questions about what their reactions meant and what to do. The issue of whether and whom to tell was especially salient at these times because they didn't yet have a self-assured way of speaking about their experience or decisions. With the exception of the two who told no one, they turned to one or two confidants with whom they could share very personal information. In every case of such sharing, those interviewed reported a positive, affirming response. As we saw with Jenna, they felt supported and got helpful advice. This benefit extended, in the relevant cases, to later sharing about professional help-seeking and medication. This support was present even in the two cases where the confidant tried (unsuccessfully) to dissuade the participant from taking medication.

As I began discussing in the last chapter, participants defined predicaments, diagnostic categories, and medication-taking in their own subjective terms (though they themselves did not necessarily think of it that way). They repeatedly used phrases such as "I just think of myself," or "more extreme than I want it to sound," or "I don't think about it that way," or "if they saw it the way I did," and the like. There was a strong emphasis on maintaining control over the language. Ryan, twenty-four, and a member of a touring rock band, gives a sense for the general idea. He felt that "sometimes I suppose I'm depressed," but he doesn't usually use that language and has never sought professional help. "I'm sure," he says, "a lot of myself would be at stake if I talked to someone about, 'Yeah, I don't know if my mental health is good.'" Attempts to describe his feelings, he feared, would be misconstrued, and people would be unable to separate the problem from the person, and so it would become "a big thing." He didn't want to be lumped in with those who lack control of their life. Rather, he wanted to define the problem in his own terms, within his own story—"the story," as he put it, "that you choose to live in."

"Witnesses" to the participants' emerging narrative, as I noted in the last chapter, were of crucial importance. These witnesses, whether confidants or professionals, contributed to and supported participants' emerging understanding, helped quell their doubts and resistance, and conferred on their views some external validation. They were "there for" them, part of their team, and this approval helped ease their suffering and sense of isolation. The subjective gained a kind of objectivity.

But talking with others could also put one's personal story in jeopardy. Most people were concerned to carefully manage whom, how, and when they told. The primary reason they gave for restricting disclosure was seldom a concern with being directly ostracized. Rather, they felt, to quote

Barb, sixty-five years old, recently widowed and retired, "that most people just wouldn't understand." In context, this concern typically had a specific meaning. Others wouldn't understand how the participant was bringing the disparate elements into an order that reestablished his or her agency and control because they wouldn't understand the predicament, diagnosis, or medication-taking in the terms the person had chosen. They wouldn't, in short, affirm the story as the person told it. This prospect was deeply troubling; self-definition was at stake.[39]

To affirming witnesses, we have to add another category, those I will call "neutral" and "disaffirming" auditors. These are listeners (auditors), who upon hearing the person's developing account do not, in the person's view, actively support it. They might take no stand, remaining neutral, or they might question or even contest the story and so be disaffirming. Participants themselves regarded neutrality as disaffirming—all the more so if they felt the response was actually antagonistic or demeaning—because they felt they had a right to have their story honored. So, by "neutral" I refer only to the lack of direct endorsement, not to how the storyteller understood it. Like witnesses, neutral and disaffirming auditors might be professionals, confidants, or others who hear disclosures.

Only one participant reported any direct experience of stigmatizing interactions, and the exact circumstances in his case were quite unclear. When a problem or "undesired differentness" is not immediately visible to other people, it is uncommon to have such encounters.[40] But a number of people referenced what they took to be common negative attitudes toward diagnosis or medication held by society or in their particular family—or even, in some cases, held by themselves—and gave this as a reason why they carefully restricted disclosure. Chrystal, for example, twenty-seven, is a graduate student in social work. She has been diagnosed with depression and has taken medication on and off for nearly ten years. While she regards her drug use "as just medicine, like if I was diabetic or used insulin," she believes others hold negative views. They don't "recognize depression as a real thing," she remarks, "and they feel that you can snap out of it and it's an excuse." Chrystal is also afraid that others will think of her as less stable or less reliable and is apprehensive about people knowing she takes medication: "As soon as I feel better [and go off the medication], I kind of get rid of those—that evidence, almost. I don't want people to know."[41]

Some participants, as briefly mentioned in the last chapter, did not receive the response from a professional that they expected. Nia, for example, appropriated depression independent of professional contact. After her first sessions with a psychiatrist, she was told that basically "nothing was wrong." Like a few others who had this experience, she was somewhat embittered by it and changed doctors. With respect to choosing confidants, people suggested that they restricted disclosure to those who they could be fairly certain would affirm and help them in their emerging story. They segregated others, who might have different opinions about what problems, diagnostic labels, and medication-taking signify. In the two cases where participants had not told even a single confidant, they expressed uncertainty that there was anyone who would give them such affirmation.

Participants identified two different ways that others might get the wrong idea if told. On the one hand, people might respond to disclosure by overestimating their problem and look at them "differently"—as "weird" or "crazy," or "messed up." Some thought that people commonly regard categories like depression, or treatments like taking an antidepressant, to only be for those with very serious disorders. They feared, in a sense, being misclassified. More generally and less dramatically, participants worried that disclosure would make their problem "sound more extreme than I want it to sound," in Ryan's words. Barb, the recent retiree who thought most people wouldn't understand, said this about telling some auditors: "I don't think they would necessarily jump up and down on me. It would be more like, 'Oh, that's kind of weird,'" though she also adds, "there's a lot of others who wouldn't consider it weird, so."

By contrast, a few people suspected that others would respond to disclosure by underestimating their predicament, by telling them that they didn't have a real problem and just needed to buck up. Luis, for instance, is twenty-three and working at a media company. He reports ongoing problems with distractibility and disorganization. He has never sought any medical help and has received mixed feedback from confidants, with some friends suggesting he seek help, but his girlfriend advising against it. He is very reluctant to share his experience with a professional because he is afraid that he or she might tell him he doesn't actually have a problem. A professional might tell him, "you just don't pay attention." This view, which he suspects himself, would upset him and only lead to more self-questioning.

The concern with self-questioning points to a deeper issue that ran through many participants' comments about a potential misalignment of views. They were concerned with the interpretive challenge of a neutral or disaffirming auditor. While witnesses helped reduce some of the ambiguity that attended the predicament, category appropriation, and medication use, a response of neutrality or disaffirmation heightened it. In this sense, the keenest worry about disclosure—especially at a point of crisis or a moment of decision, when the emerging account was most precarious—was that the other person would unsettle their story by doubting or questioning its validity. The unsettling, as we saw above, could be either actual or anticipated. Participants felt that disaffirming responses, such as an alternative reading of things, whether by a confidant or professional, were stigmatizing. They expected that their version should be recognized and affirmed. In other cases, neutral auditors were a concern even before anything was said or done; the possibility of misalignment was enough to inhibit disclosure.

This concern was not just a matter of other people's reactions; it sometimes originated with the participants themselves. A number of people worried that talk of their predicament could become a self-fulfilling prophecy.[42] Michael, for instance, a recent college graduate, experienced anxiety in social situations that he did not think warranted any sort of treatment. He is the young man quoted above who remarks that having a disorder would not make you a bad person but might make you a "different person." When asked about disclosure to others, he explains his avoidance not in terms of others' reactions but in terms of his own desire to regard himself as normal. For instance, in the interview, we asked if he had ever considered seeking help, such as attending a self-help group. He answers: "I don't want to say I'm like that [like those who do attend self-help groups] just because I don't think I am." Then, he adds, "maybe it's more like I don't want to be, more than I don't think I am." When he concludes with "I don't want to be considered messed up," it seems clear that what he was talking about in the first instance was his own regard for himself. He did not want to think of himself as messed up, and telling others would make that more difficult. It would somehow reinforce or "validate," to use another person's word, that he was having a hard time. Many people spoke in this way, combining a concern with others' reactions with a concern about their own.

The same dynamic was at work in the concern that a few mentioned about how disclosure might draw from others a response of sympathy or pity. At issue was not so much a concern with misunderstanding as with embarrassment at falling short of expectations about self-efficacy, about what one should be strong enough to handle on one's own. These were the only people who directly spoke of feeling ashamed by their situation (though it was certainly implicit in others' remarks). They indicated that telling others, either about the suffering itself or the need for help, would make matters worse. The clearest case was Amber, twenty-seven years old and struggling emotionally after her boyfriend left her and their two children. "I'd rather not have sympathy," she says, because I feel like that would be weak. . . . When somebody else is feeling bad for you . . . you feel just a little more inferior if they're looking down, and I feel like I have to be stronger than that, or like we're equals still, even though I'm going through this.

For these people, sympathy is a response to vulnerability or needing help, and drawing such a reaction, however well intentioned, would both diminish them in the eyes of others and cause them to see themselves as "weak" or "inferior." That potential response was a powerful reason to carefully protect what of their experience they exposed to others.

A final and commonly expressed reason for nondisclosure also seemed to have more to do with participants' views of themselves than with the attitudes of others toward them. Some simply described themselves as private persons. They were not ashamed by their situation but saw no reason why they should speak to anyone about their predicament beyond their small circle of confidants. That was the reason given by Georgia, the woman we met in chapter 3 who was diagnosed with depression during a difficult period of financial belt-tightening. But she was also conscious of engaging in impression management that was, in part, for the sake of her own self-understanding as a person with strong willpower. The movement to regain control began with projecting a positive image to herself, while maintaining that appearance with others.

## Diagnosis: Between Normality and Mental Illness

With regular medical diagnoses and treatments, like diabetes and insulin, where clear and common understandings of malfunctioning exist, personal definitions are of little importance. But in the case of participant predicaments, the personal meanings are pivotal. The very impression point came when a definition of the situation or a course of action emerged that either provided or accommodated these meanings. Although some participants said they didn't think their definitions would necessarily be shared by their doctors, such explicit contrasts were unusual. Their struggles were not with health professionals, who neither imposed on them nor required of them any specific understanding of their predicament, diagnosis, or medication. Practical decision-making in the clinic required compliance, not shared meaning. Their struggles, rather, were with those incompatible aspects of their situation that tarnished their agency, sense of freedom, and self-understanding. In a social environment in which everyday suffering has been heavily medicalized, part of this struggle, as we saw with Jenna, involved finding a way to understand and describe their predicament as tangible and unfeigned but without some of the uncongenial and undesirable implications of medicine. With respect to diagnosis, the key implication was "mental illness" or "psychiatric disorder" and its frightening connotations of differentness and loss of control.

### *"Not to the Extreme of an Illness"*

Both for those who appropriated a diagnosis and those who did not, diagnostic categories were important for defining predicaments. As noted in chapter 3, the people who did not identify with a diagnostic category were, with just two exceptions, psychologizing in their explanatory perspective. They had not sought any medical help or been diagnosed by a physician and, at the time of the interview, did not intend to pursue such help. (A few, however, explicitly left open the possibility that they might decide differently if their circumstances changed or their life did not improve.) They defined their suffering as of a different kind than what they took to be a diagnosable psychiatric disorder. Although some spoke in a way that suggested that they viewed mental illness on a spectrum with normality, on closer inspection their remarks generally revealed a more categorical view. Mental illness was a somatic disorder, and they used such disorder as a point of contrast in describing their own experience.

Many in this group suggested that the sign that their experience is "not to the extreme that it's an illness" is that a diagnosable mental illness would require "some kind of pharmaceutical intervention" and their case does not. These quotes come from Phyllis, sixty-one, single, and retired, who struggles with shyness and fear of criticism and who characterizes her periodic "funks" as depression. She has seen therapists at various times in her life and has been formally diagnosed with GAD. But she does not accept the diagnosis. She regards it as a mere insurance necessity: her health providers, she says, told her "kind of apologetically" that they had to record a diagnosis for the sake of insurance reimbursement. One of the reasons, she explains, is that those who have an illness could be greatly helped by medication and, while she thinks medication might be of some help to her too, she does not need it. Hers is more of a "personality thing," and unlike in the case of a mental illness, her psychotherapy has helped her to cope.

People in this more psychologizing group did not regard their suffering as normal, but they also did not see it (or want to see it) as making them different from other people. They stressed personal effort and were typically resistant to medication because they believed that it would be an unacceptable crutch or excuse *for them*—though others, worse off than themselves, might appropriately resort to it. Their predicament, at least at the present time, did not warrant the use of a drug. Some simply hoped the predicament would resolve itself. These participants also typically expressed more willingness to live with, if not fully to accept, limitations.

### Finessing the Illness Implication

Control over defining their predicament was one thing for those who had not come under medical management; it was different and more challenging for those who had. The DSM diagnostic categories are forms of disorder/illness/syndrome/disease, and this language can be threatening. We have already encountered the many ways in which participants distanced themselves from the formal categories ("that's what the doctor calls it," "I just call it 'anxiety,'" etc.) or qualified their import (the doctor "thought I was kind of depressed," for instance, or the doctor "said I had a slight ADHD"). Further, as many observed, talk of mental illness conjures up images of very serious problems: "someone who's babbling to themselves," according to one. Other studies have found that even those with fairly severe impairment tend to avoid the language of mental illness. For instance, in sociologist David Karp's interviews with people suffering from serious depression (e.g., three-fifths had been hospitalized at least once), he found that only a few "were clearly willing to define their condition as a mental illness." Most sought, to quote one of his participants, to "finesse" the "illness thing." And these were people, Karp reports, who adopted a conventional "medical version of reality."[43]

It is no surprise here, then, that fewer than one in five people who appropriated a diagnostic category allowed that their suffering might be described in a language of illness. Those who did seemed to share the view that medication is for illness, and so if they took medication, then, at least technically, there must be an illness present. But the illness, they also emphasized, did not involve any significant impairment or loss of control. On this point, several made an analogy with diabetes or coronary risk factors. This was a way of saying, according to Chrystal, the twenty-seven-year-old social work graduate student mentioned above, that her diagnosed condition (depression) and medication-taking were very common medical phenomena and not especially serious.

The other people who appropriated a diagnostic category took a number of different steps to distance themselves even further from the implications of mental illness. A number of those who interpreted their predicament from a psychologizing perspective had been formally diagnosed and were taking a medication. Their appropriation of the diagnosis seems to have followed, and been effectively necessitated by, their medication-taking. They defined their predicament as a troubled response to a situational dilemma and, in several cases, insisted that their medication use would be temporary; they were not, they emphasized, suffering from a chronic condition.

Some of these people felt that their category would only be a mental illness if they chose to let it become one. In the aftermath of her second divorce, Melissa, fifty-four and a professional illustrator, was having trouble sleeping, crying a lot, and had little energy. She went to her family doctor, was diagnosed with depression, and was prescribed an antidepressant. While acknowledging her diagnosis and agreeing that it's termed a mental illness, she believes that viewing depression as a disease gives it too much, self-fulfilling power: "if you end up saying it's a disease," she says, "I think that gives the disease power over you rather than you power over it. So, depression, for me, is the same thing." She also emphasizes the importance of a proactive response to medication use. For these participants, a diagnosed condition could sometimes be considered a mental illness, if you let it dominate you, but their particular case was different.

In distancing themselves from the idea of mental illness, these people, like those who passed on a diagnostic label, were also distancing themselves from the idea that their problems had a somatic origin.[44] They didn't have a problem, each indicated, with their "mental faculties." From their remarks, it was clear that having an illness caused by some biological malfunction, far from easing the threat of mental health problems, was the gravest challenge. It meant a loss of volitional control and a causal determination of their actions and emotions; it meant they were different and unfree. In their appropriation, they rejected this difference. If taking medication meant that they could not wholly separate themselves from a diagnostic label, they qualified the meaning of the label by separating it from mental illness.

The challenge of the somatic and the specter of determinism was even more sharply felt by those who appropriated a diagnostic category with a medicalizing explanation, whether taking a medication or not. While some of these, as noted above, spoke in terms of illness, the balance sought to separate their suffering from mental illness, even as they insisted that it was caused, wholly or in part, by a physical malfunction. They accomplished this segregation by positioning their suffering as a kind of "third condition" (my term)—neither normality nor illness exactly, but something in between. In a number of cases, they credited a medical professional or confidant for suggesting something like this idea to them: emphasizing that their suffering was common—many ordinary people, they were told, go through the very same thing—and that it could be (seemingly easily) managed, even "fixed." This framing conceded an unavoidable element of differentness, while also containing it.

People did not define their "third condition" directly, in terms of what it is. They were engaged in personal meaning-making not some larger theoretical exercise. Besides, the point of this framing was not so much to introduce a new type of differentness as it was to mitigate existing implications of differentness. The third condition was a space between the things it was not, not normal and not mental illness. In fact, it might be accurate to say that this space was constituted on the claim that whatever the participant had, it did not have a predefined meaning. With a "third condition," the participants were not under the control of the disorder (category). It did not define their experience; they did.

The clearest example came in the way that some participants drew a sharp distinction between their experience or appropriated category and serious mental illnesses, such as schizophrenia or bipolar disorder. We have already seen examples of this kind of splitting, such as in the story of Eric (chapter 2), the teacher who was overwhelmed by the pressure at his "high-rent" school. He distinguished his "anxiety"—his word, used instead of the formally diagnosed GAD—from conditions like schizophrenia that he called "hard-core." In this distinction, people not only claimed that their own experience was a less serious case, they also suggested that it was not a "condition" in anything like the same sense. It was of a qualitatively different kind, its meaning distinguished in part by not having any fixed meaning. Schizophrenia or bipolar disorder, they suggested, was not something you personally appropriate.[45] Much like a regular physical illness, the meaning of such serious disorders is outside the sufferer's personal control. Their experience, by contrast, is not determined in that way.

Alternatively, but to the same purpose, some compared their experience to that of other people whose circumstances they felt resonated with their own. We saw several examples in the last chapter. In such cases, participants stressed the closeness of their experience to that of another (or to an image in an advertisement), whose struggles modeled for them issues that were the object of medical attention (diagnosis) and treatment but that did not seem like a mental illness. These problems did not, for example, appear to involve a substantial loss of control or create a dependence on doctors. Comparison to the model offered them a middle ground, at a safe

distance from mental illness but not quite normal either. Their shared diagnosis was a "witness" to the facticity of the suffering, confirmed the biological explanation, and warranted the medication treatment. They had a
third condition.

In another, overlapping pattern, participants framed a third condition
by adopting other terms with different meanings to draw the contrast to
both illness and normality. They too saw their experience as qualitatively
different from a mental illness. Piper, for instance, says, "you think mental
illness you think schizophrenia and crazy people and I'm not crazy, [I] just
get really nervous." And malaria is a disease, she remarks, "not I'm sad."
Rather, Piper thinks she has something else, which is a further reason these
participants give for why mental illness is the wrong category. She observes,

> I would say more like you have just a condition, like a slight—something is a
> little off. You have a gene that's a little messed up that's maybe not function-
> ing as well as other people's are. . . . I think of it more as, some people have
> blue eyes, some people have green eyes, some people have brown eyes, some
> people have too much of certain chemicals, some people have too little,
> some people have perfect, and I'm just—I think I just have too little or too
> much or whatever it is that makes you have these issues.

Piper has something different from an illness, which she is unsure how
precisely to name. At the same time, it is third in the sense that it is also
different from the normal vicissitudes of life. When identifying the cause of
her suffering as an "imbalance," Piper is equally insistent that her anxiety
in social situations is different from ordinary shyness.

## Medication/Psychotherapy

In their efforts to overcome their predicament, the majority of people
treated "mental illness" as a threat to their self-image. It implied a loss of
control, a differentness from other people, and an inability to cope on one's
own. These were hazardous elements that participants sought to mitigate
or avoid. For most, mental illness also implied a need to be medicated.
Interestingly, as noted, those who were not under medical management
were typically the most explicit about what they saw as the close, seemingly
necessary connection between diagnosis and drug. It was often one of the
reasons why they did not seek medical help. Maya, who is deeply unhappy
but refuses to see a professional, is perhaps the most succinct and forth-
right: "I don't want to be clinically depressed because that means that I'll

need drugs or something and that is horrible." But nearly everyone made the connection.[46]

Medication, too, despite its ubiquity and reputation for efficacy, could be spoiling of self-image. Without question, most participants who settled on a medication (a fluid process with considerable brand switching and experimenting with dosage levels) regarded it as helpful in some way. They reported a number of distinct personal improvements and identified important ways in which their medication experience had influenced their experience of self. I will discuss their observations in more detail in chapter 6. But for most, medication use also carried ambiguities. There was an expressed tension between its benefits and disquiet about what pill-taking means. Clearly, there was a social norm at work: drug-taking had to be justified and was only appropriate if, and for as long as, there was a real problem and the problem could not be personally controlled. In light of such a norm, some participants expressed concerns about weakness, dependency, and loss of control, and many others worried about a kind of personal "falseness" brought on by medication use. These implications had to be dealt with as well.

Some people expressed a direct concern with dependency. A few worried that taking medication might create a *physical* dependence. Their response was to do things like maintaining a low dose or otherwise regulating or cutting back on the amount they took. More commonly, participants expressed concerns that taking medication meant a loss of independence and self-sufficiency. Will, for example, forty, is working as a freelance web designer. He sought help for anxiety problems at work and loss of energy and desire to do things. He was formally diagnosed with "anxiety" and "major depression" and had been taking a medication for each condition for two years. He had also been in counseling for six months at the time of the interview but felt helpless to "fight through" problems on his own and had little choice but to take medication. Will emphasizes, though, that it makes him "feel inferior." Taking medication, he says, means "just dependency. I just don't feel in control when I'm kind of living life on pills." He has avoided telling anyone except a previous girlfriend about his diagnosis and medication use.

Many people also expressed a concern that their use of medication might be inducing some personal "falseness." This concern took several forms. One was a worry about enhancement: the idea, in essence, that their medication might be too effective. Rather than easing the symptoms of a medical condition—"leveling the playing field," as some put this idea—it might be conferring some unjustified advantage. They might be "using it," according to one participant, "to get a step up on everyone." This concern was clearest in cases where the participant was using a stimulant medication, like Adderall or Ritalin. A great deal has been written about the use of such drugs for purposes other than treating disorder, most notably their nonprescribed use by college students as an aid to study.[47] Participants negotiating the meaning of ADHD and the use of stimulant medication were well aware that, as with doping in sports, drug use to enhance performance is often viewed as a kind of cheating. And some, although formally diagnosed, were apprehensive that they might be gaining some illegitimate advantage.[48]

Another overlapping concern was with the common criticism that medication use constitutes an "easy way out" or inappropriate "quick fix." This criticism was only occasionally directly expressed. Monique, for instance, fifty-six, had taken an antidepressant in the past and found it helpful. She was even considering starting again. But she thought medication often served to camouflage or mask real problems and believed that people were often motivated to take it because it allowed them to avoid facing the underlying reasons for their difficulties. Generally, though, the concern that using medication simulated improvement while sidestepping appropriate self-work was not so much expressed as suggested by other things that participants emphasized. Some, like Jenna, or Georgia in chapter 3, stressed their proactive engagement with the necessary medication. And others, like Lisa from chapter 1, emphasized all the things they were doing besides taking a drug. They were not avoiding the hard road; they were taking it.

A final form of commonly expressed falseness concerned ways in which drugs were unnatural, blunting feelings too much or effecting undesired changes in personality. Odetta, for example, forty-seven, who works in a financial firm, describes her formally diagnosed depression as arising from a "situation." She has dealt with grief over the death of her adult son and later her (estranged) husband and with the painful feeling that her life has been a failure. She sees a psychiatrist for psychotherapy and also takes an antidepressant. The medication blunts her feelings, which, on the one hand, she credits with keeping her emotions from being "all over the place." On the other hand, it causes her to worry. "Maybe it is unnatural," she says, "and even though feeling extremely bad or sad or whatever, it's not a good thing, at least it's a natural thing." For Odetta, the bad feelings might be truer.

In contrast to medication, most people seemed to regard psychotherapy, in the words of one, as "not an embarrassing thing." The exceptions, as noted in chapter 1, were primarily a few men, who did express resistance to the idea or reservations about what others would think. Clearly, however, though participants said little, much of what I have said about controlling disclosure and worry about disconfirming others might apply to reluctance to see a therapist. As Jenna made clear about her reticence to contact student health, all disclosure involves apprehension. It means acknowledging a problem and opens the possibility of undesirable "ways this could go." At the same time, as Jenna indicated, going on a medication, compared to therapy, was "a much bigger step." Therapy or counseling might be very short-term, involve no diagnosis, and carry no implications of mental illness or dependency or falseness. Much less threatening in that.

## Conclusion

At the heart of everyday suffering is anomalous experience, and participant sense-making involved positive steps to negotiate threatening elements and establish right order. For many, medical help-seeking and treatment presented a clear danger to self-image, whether in disclosure, diagnosis, medicalizing explanation, or medication. The threat was not so much from a fear of negative stereotypes as from a gap between the goal of viable selfhood and the means to achieve it. To shrink that gap—to neutralize the threat of differentness—participants shaped definitions and made interpretations consistent with the positive picture of self toward which they aimed. In most stories, the goal that emerges is a picture of themselves as proactive, in control, and especially, self-defining.

This goal comes to light in the values that participants emphasize. They made their own decisions, in consultation with others but on their own terms and not in mere deference to the judgment of medical professionals, family members, or friends. They stressed that their remedial actions included personal effort, an application or expression of their control. Medication, for instance, was not "magic," or a "quick fix," or an "easy way out." They were responsibly engaged in appropriate struggle and concerned and vigilant to avoid dependency.

The desired self also emerges in the background norms and ideals that participant evaluations must presuppose. Recall again Jenna and the situation she faced after the cheating by and breakup with her boyfriend. She says remarkably little about her emotions. "Sad," "crying over nothing," "tired," and "depressed" are virtually the only negative emotion or emotion-related words that arise in the entire interview. She makes no mention of complex, social emotions such as jealousy, betrayal, anger, embarrassment, or resentment that one could reasonably expect her to feel in the circumstances. Instead, she characterizes her (generic) bad feelings as unreasonable and without meaning, not her emotions at all but emanations from a source outside herself. She draws on social norms of emotional efficiency and control to characterize this experience as anomalous and indicate what her good self, if unimpaired, would have felt and done.

Attention to the (often implicit) social norms and ideals by which participants evaluate their selfhood helps us to see how they gauge what is abnormal in their experience and what reestablishing order appears to require. Dealing with the dangers surrounding medical help-seeking was one window on the broader effort to put things right. In the next chapter, I continue the reflection on what makes the neurobiological compelling by further exploring the normative picture of viable selfhood toward which most participants aimed. Illuminating this regime of what it is good to be also suggests the terms in which it might be realizable. Those are terms, the possibilities for self-definition and normative performance, for which the neurobiological imaginary is uniquely accommodating.

# FIVE

# Seeking Viable Selfhood

Rob, forty-one, has a history of painful breakups. It began, he recounts, years earlier with the first woman he dated seriously. Just when he thought they might have a future together, he discovered that she had cheated on him with another man. He took it very hard. He felt "out of it," unable to focus on other things, and experiencing painful feelings of emptiness and self-criticism. When these emotions didn't subsequently subside, he grew concerned that something was wrong with him and that he needed help. In the assessment of the psychologist he saw, however, there was nothing really amiss, and eventually the feelings did decrease and cease to trouble him.

A few years later, Rob again experienced an unanticipated rejection by someone he really cared about. All the strong feelings came back, leaving him again feeling hurt and empty. He started questioning everything in his life, puzzling over the intensity of his responses and inability to just move on. He returned to the psychologist, who this time conducted some brief therapy with him. In Rob's view, "it just didn't seem to really do the trick." In time, though, he met someone new.

After yet a third breakup and persisting emotional turbulence, Rob was desperate for answers and anxious to deal "once and for all" with whatever was going on. He went back to the psychologist, who now suggested that he see a psychiatrist for a consultation. He was tired of feeling like an emotional failure, and again thought it was abnormal to feel so depressed by the breakups, obsessing over them and losing sleep. His friends, he says, thought so too.

In telling his story, Rob stresses the hurt of his romantic relationships, but his work life has also been a recurrent sore spot. Over the years, he has held many different jobs and has been consistently unhappy in them, whether it's the people, the pay, or the work itself. These work troubles have caused him much anger and frustration, reactions he also sees as odd and disproportionate. His friends also affirm his view that he overreacts. They believe he shouldn't get so disappointed and remain upset. They think he needs a thicker skin, a view that irritates Rob, who insists that he will be the judge of what he does and doesn't need.

Rob took the advice to see a psychiatrist. The consultation, he reports, was basically a question-and-answer session. In the end, the psychiatrist, observing that the psychotherapy didn't seem to work for Rob, diagnosed him with a "middle-level depression" and wrote a prescription for an antidepressant.

Initially, Rob was concerned about the drug, worried that it would change him into a different person, even "like a robot." So, he sought out a friend who he knew had been taking an antidepressant for some years. The friend, who was a practicing attorney, told Rob how the medication had helped her. After this conversation, he made up his mind and agreed to try the medication. In about two weeks time, he started feeling better. His emotions were more "level," and he was able to concentrate a little better.

In Rob's view, "depression" is not a good word for his experience. During his "phases," as he calls them, he feels down some of the time but more energetic at other times. When he and the psychiatrist discussed some troubles he had as a teenager, it was shyness, not sadness or disappointment, that had given him trouble. When he received the diagnosis, he was actually "kind of bummed out" because having depression carries public stigma.

An impression point came with a neurobiological explanation. Referring to his psychiatric visit, he recalls,

> I don't remember him issuing a theory. I think he kind of explained to me that there are differing opinions on what causes it [depression] and what doesn't, but he . . . explained the neuroreceptors with the serotonin thing in your brain sometimes, and I just thought, well, maybe that's it.

Beginning with this suggestion, Rob sought information from other sources and has renarrated his past as an unrecognized history of depression. He now sees himself suffering a lifelong affliction, genetic in origin, which accounts for his intense emotional reactions to his breakups and job problems as an adult and his shyness as a teenager.

In the new account, Rob sees his unruly emotions as foreign to him, without any intentional object. They are the result of a physical condition that has kept him wallowing in the past. Without this condition, he argues, "[if] I didn't have a job or I didn't have a good job or didn't have a stable relationship, I probably would change those things without a second thought. I wouldn't even think about it. Or I would just cope."

For two years, he felt the medication was working, but then he lost his health insurance and had to stop. When he was finally able to resume treatment, the drug didn't seem to work as well, even when combined with other antidepressants. Since then, he hasn't found anything that consistently makes him feel better, though he has a clear idea of what he would like. An ideal medication would allow him to decisively move on, handling all of life's situations in a way that didn't leave him feeling either guilty or stuck. Without a medication that can produce that ideal and convinced psychotherapy will not help, Rob has become fatalistic and wonders if he will ever feel better.

---

For Rob, like Jenna, the young woman discussed in the last chapter who was also struggling with emotions occasioned by infidelity and a painful breakup, the reported predicament was not the relationship loss or feelings of betrayal or abandonment per se. At least initially—at time 1, so to speak—he felt some such emotion, presumably grief, or rejection, or disappointment, and saw it as a meaningful response to the situation. Like Jenna, Rob actually says little about the initial emotions he was experiencing. He refers to sensations of feeling "really weird" and an "emptiness" and is surprised and distressed by their intensity. Even more, he is confused when they persist. At time 2, he construes his feelings of being depressed and frustrated as inappropriate. He does not regard his circumstances now as warranting any strong feelings or their continuation.

These judgments certainly reflected the fact that Rob felt himself to be in a rut and his rumination and inability to let go of past hurts was causing him considerable distress. But the very definition of his predicament presupposed a standard external to himself, one that he implicitly referenced with the witness of his friends. They confirmed that his time 2 emotions (again, unspecified) were anomalous. "Is that all?" he reported them saying with respect to the breakups and unsatisfying jobs, "well, there's other fish in the sea, other jobs." He marks his emotions as "odd" by comparison to a common emotional standard, shared by himself and his friends. He *should* not be feeling as he does; he *should* have moved on in short order. That is what nonimpaired people, apparently without struggle, would do.

As Rob's story shows, predicaments presuppose social norms—whether they be moral tenets, legitimate aspirations, acceptable ways of living, valued status attributes, performance expectations, or other types of obligations, standards, and ideals about relationship, emotion, and self/being.[1] We have encountered such norms consistently, most directly in the last chapter. In defining and dealing with their predicament, diagnosis, or medication, participants expressed concerns about a loss of personal control, insufficient effort, dependence, falseness, inappropriate emotions, and more. All these assessments have an inescapably comparative dimension.[2] They make no sense in any other terms. Judgments of what is acceptable or unacceptable, better or worse, praiseworthy or blameworthy necessarily come into play that are both independent of us (they are not the invention of individuals) and constitute standards against which our conduct, thought, and feelings can be understood and appraised.[3]

Many norms that govern everyday conduct are merely conventional categories. The norms at stake here, social norms, are much more than that. They concern our very status as persons in society and always reflect tacit value systems about the good. While there are certainly degrees of prescriptive force and institutionalization, social norms carry an "ought" that gives them their power or force in our lives. In sociologist Margaret Archer's words, social norms "both categorise our actions and attach evaluative judgments to them," representing some actions, states, or relationships to us as being offensive, unacceptable, or morally reprehensible, others as good and right and appropriate.[4]

Although social standards and ideals are external to us, how we evaluate ourselves with respect to them has an inherently subjective component. In a complex society like the United States, there are multiple normative registers, and even deeply institutionalized norms do not appear binding or convincing to everyone. There is tension and conflict between norms, and we can critique or reject or be indifferent to particular norms. "It is perfectly possible," for example, as Archer continues, "to be wholly indifferent about school achievement, whilst dispassionately recognizing the standards and expectations involved."[5] For social norms, unlike, say, legal norms, to really affect us, we have to care about them, to have internalized some sense of the standard.

Social norms, especially in the linked matters of the good and subjectivity, often remain taken for granted and in the background. They may only come to awareness when challenged or violated, when there is some deviation or failure, such as in everyday suffering. This means that people might not be conscious of them. In describing his predicament, for instance, Rob was to some degree aware that he was drawing upon social meanings of normal, while also regarding his self-evaluations as reflecting his own personal and private judgments, free of social influence. Other participants said similar things, sometimes recognizing and articulating a social standard but often not.[6]

To see what evaluation participants were making, I often had to infer what expectation or standard or ideal they must be using for their statements or the emotions they expressed to make sense.[7] This was to be expected. Their evaluations and comparisons do not necessarily represent any systematic philosophy or comprehensive vision of the good to which they consciously subscribe. For most of us, to quote anthropologist Mary Douglas, "Our view of the world is arrived at piecemeal, in response to particular practical problems."[8] While everyday suffering is a unique window on our subjectivity because it touches so directly on our self-worth, we would not normally speak of a single orienting sense or principle—such as autonomy—but of a multiplicity of context-relevant goods and values. Except in those cases in which people are knowingly shifting normative registers, they typically draw upon meanings that just *are* what seem appropriate or inappropriate, acceptable or unacceptable, right or wrong, without reference to their source or justification.[9] Comparisons to norms are implied in the stories, and so can be deduced, though they often remain unspoken, unrealized, or unacknowledged.

In this chapter, I explore in more detail the social norms of relationship, emotion, and self against which participants judged themselves and defined predicament resolution. Drawing out the relevant norms of being is critical because how participants understand and respond to them has implications for how they imagined what lies behind their predicament, why it happened to them, what establishing viability requires, and how the outcomes of their efforts, including the effects of medications, are understood and evaluated. It is precisely here, in the matter of what the good self requires, that we see what makes the neurobiological imaginary—an imagining of how things are and how they *should be*—compelling.

### A Common Normativity

As discussed in the introduction, the specific circumstances and sources of discontent participants identified as relevant or contributing to their predicament clustered into the broad categories of performance, achievement, and loss. In an obvious sense, these struggles are different from each other. They are defined against somewhat different standards and typically involved (if they involved at all) different diagnostic categories. Problems with social performance, for instance, presupposed some standard of out-goingness and extroversion, and if a diagnosis (whether formal or just appropriated) was involved, social anxiety disorder was the most common. Similarly, problems with academic underperformance were typically measured against standards of concentration and organization, and ADHD was the most common diagnostic category. And so on with the other types of predicament.

In contrast, when we consider the terms in which participants envisioned amelioration, we find a different pattern. Regardless of the specific circumstances and discontents, restoring right order entailed efforts to conform to quite similar and related norms of being. Being a person who is proactive and optimizing, for instance, was a standard for many participants dealing with underperformance, but the same social norm shows up for participants dealing with achievement and loss. The same was true for the other social norms described below—dealing with emotional efficiency, personal autonomy, and self-approval. The most salient differences between most participants—at least as could be teased out within the limits of my method and a single interview—were matters of emphasis or articulation with respect to their circumstances rather than of different standards. There were some clear exceptions, especially along social class lines. Not everyone shared the same values, much less to the same degree; I do not mean to suggest a uniformity. But the similarity, the family resemblance, stands out and a broadly shared normativity to which it points.[10] This is the sense in which we can speak, in an Aristotelian sense, of a "regime," a coherent vision of the good that governs and orders us.

The emphasis in participant stories was less on where they have been than on where they are now and where they hope (whether by choice or necessity) to go. The crucial norms against which they judged themselves, whether explicitly or implicitly, spoke to the end state they desire—the return of their self-respect, for example, or the reassertion of their control, or getting on with a more productive life free of bad feelings. For most people, the aspiration was not simply to resolve the predicament but to reconceptualize themselves in such a way that they lived more "viably" and predicaments like the one they were in might not arise again.

## Viable Selfhood

The normative standards, including the social norms of how to conduct relationships and control and express emotions, all concerned selfhood—

subjectivity, one's relationship to oneself—and the sort of person one is. In considering how participants negotiated medical language and implications, I sought to show that it involved reconciling interpretations with an idea of themselves as "viable" selves. The people we interviewed did not actually use the term "viable" to appraise their relationship to themselves. According to the Merriam-Webster dictionary, one of the meanings of "viable" is "capable of working, functioning, or developing adequately," and, in another variation, "having a reasonable chance of success." While this is not a usual way of talking about persons, I think it fairly represents the evaluative framework that came across in many participant stories, whether with a medicalizing or psychologizing perspective. As their quoted words have already demonstrated, they often used a thin, instrumental language—functional, efficient, rational, productive, or inadequate, imbalanced, insufficient—and strongly emphasized and valued a present-oriented and forward movement: the ability to carry on and not get stuck or hampered by past events or feelings. For many, ameliorating or rising above their predicament was not a matter of, say, healing or working through, but of a mechanical-like "fixing." And they often stated the end point or goal as being or feeling more like they should and more like other (viable) people do. Life is a project that can be measured in terms of success or failure. "Viable" seemed like an apt word to capture this thinly conceived yet consequential view.

I have argued that what people are striving for is living up to an ethical picture of what it is good to be. "Viable" isn't a moral word, and participants did not use a language of ethics, or morality (in the sense of moral codes or repentable transgressions), or ideals of the good. On the contrary, I think it is fair to say that most did not regard themselves as engaged in anything like ethical reflection. They would not regard how they think about themselves to be an ethical matter. In liberal society a certain denial of the normative nature of human life is part of the dominant normative picture itself.[11] Autonomous selfhood is conceived as prior to moral order, so, to quote the philosopher Iris Murdoch, "the 'inner life' is not to be thought of as a moral sphere."[12] That, I think it is safe to say, is the background assumption.

In order to see the ethical picture, then, I have to emphasize what is left out. In normal speech, we typically reserve the terms "moral" and "ethical" for issues of justice and respect for others, and for those matters of right and wrong, permitted or forbidden, that are explicitly codified in moral codes or statements of ethical principles. But that meaning is far too narrow. In a famous interview that he gave toward the end of his life, the French philosopher Michel Foucault suggested that in order to trace a history of morals, we have to distinguish "acts"—the actual behavior of people in relation to "moral prescriptions"—from the "moral prescriptions" themselves. Foucault did not regard official moral codes and guidelines for behavior as particularly decisive for understanding moral change. They tend to remain fairly stable over time—think of the Ten Commandments—while the substance of moral subjectivity clearly varies. He emphasized instead "another side to the moral prescriptions, which most of the time is not isolated as such but is, I think, very important: the kind of relationship you ought to have with yourself, *rapport à soi*, which I call ethics, and which determines how the individual is supposed to constitute himself as a moral subject of his own action."[13]

This formulation—"the kind of relationship you ought to have with yourself"—precisely describes what is at stake in predicaments. This relationship, our subjectivity, has an ethical and public dimension. It concerns our orientation to the good and to our own dignity and sense of purpose and meaning. It concerns our self-worth and what we have invested ourselves in, our attachments, our beliefs, the things we count as truly significant. It involves "oughts"—how we are *supposed* to constitute ourselves—and so involves moral discriminations, appeals to authority, and practical techniques and efforts to conform. To miss this ethical element is to miss what participants care about, the onus of their self-evaluations, the standards under which they are laboring, and the ideals they are struggling to achieve.

While speaking of "viable" matches a thin way of talking, what is in fact in view is a thick concept of the kind of people that participants take themselves to be and feel a moral responsibility to produce. In the exposition that follows, I will consider the common, if differently emphasized, norms of viable selfhood and briefly consider some general evidence of this normative system in American society. I will also consider a few notable exceptions. Some followed or sought to follow different norms of being.

### Relations Contractual

I return to Rob, whose evaluative outlook expresses many of the major social norms I found among participants, from self-sufficiency to emotional efficiency to personal adaptability. With respect to his breakups, Rob's predicament is keyed not to the loss of relationships as such but to his failure to let go. In the face of rejection, he has failed to be appropriately self-sufficient. He should have shrugged off relationships when they ended, no matter the reason, and moved on with his life. Same with his jobs.

Rob does not speak of norms. On the contrary, he insists that his standards are his own and that he does not answer to anyone about how he defines his life. Most participants, of course, spoke that way. A key aim was increased autonomy, which means free choice without coercion or conformity. It is hard to conceive of autonomy itself as a kind of duty, a way of conducting ourselves that is the subject of social approval or opprobrium. Yet, external standards are clearly involved. The way Rob speaks of his breakups seems to indicate a vulnerability and gullibility surrounding intimacy and commitment. Each time he was caught unawares and appears to see a symptom of his condition in the fact that he did not gauge the commitment of his girlfriends and then mirror his own accordingly. He *should* have maintained more autonomy and control.

The relationship norms at work here share a close resemblance to the norms of what sociologist Anthony Giddens calls the "pure relationship."[14] Giddens derives his ideal-typical account in part from logical deductions about the effects of social change in recent decades and from the therapeutic literature. His description represents a fair picture of the contractual ideal presented in popular self-help writing. It is an ideal that has arisen along with, and in turn fostered, the progressive evaporation of the external ordering criteria (fixed social positions, predefined social roles, traditional frameworks, etc.) that once gave close personal connections an anchor.

The key feature of the pure relationship is that it is prompted by nothing other than what it can bring to the partners involved. It is, in this sense, and unlike, say, kinship relations, something purely chosen and, therefore, involving greater risk and uncertainty. Intimacy is more precarious at the same time that it is more highly valued. It carries more weight and becomes a kind of crucible—providing an emotional anchor, personal security, the satisfaction of needs, the expression of intimate feelings, a forum for exploration. But intimacy is an achievement, and the trust that must be generated requires romantic partners or friends to reflexively and continuously monitor the relationship for mutual alignment and reciprocity in commitment and psychic satisfactions. The relationship, properly maintained, requires that each person retain their autonomy and confidence in their own self-definitions, such that they can be relied upon to respond in predictable and satisfying ways. Personal closeness is not to give way to dependence or, conversely, a taken-for-grantedness. In pure relationships, each partner is engaged in a linked process of both self-exploration and the development of intimacy with the other.

Many of the interview participants who were dealing with situations involving personal relationships implicitly judged their conduct against such contractual type standards. Like Rob, emotional fallout from a loss sparked a reevaluation of their previous conduct that was being lived by some other, not so pure relational criteria. In context, these reevaluations appeared less as the vital embrace of a positive ideal than as steps to protect self-worth in the light of personal rejection and emotional responses deemed anomalous. Either way, the norms are the standards against which experience is being measured. Rob clearly implies that if he did not have a biological disorder and therefore had been living by the appropriate social norms, his experience would have been very different. He would have quickly and easily let go, moved on emotionally, and confidentially embraced his new circumstances. Much of the suffering could and should have been avoided. The issue going forward, about which he has grown skeptical, is how to find the right medication that will allow him to be more detached.

Though common, not everyone measured themselves according to such normative criteria. Dwayne, forty-seven, for example, works in a customer service department. He describes his predicament in terms of his isolation. He had long enjoyed the support of a small group of very good friends. However, over a year-and-a-half period, he lost all of these relationships as one moved away, two died (cancer, a drug overdose), and two withdrew from regular contact because of difficult personal situations. Now he has no one to talk to. He is caring for his aging father, who is in deteriorating health, and communicates very little with his wife, whom he describes as going through "this menopause thing." He expresses remorse for not doing more to help the close friend who died of an overdose. Feeling drained and "stressed and angry," he has begun to drink more and withdraw from his old pursuits.

But what Dwayne emphasizes is not the problems as such. "Everybody," he says, "goes through peaks and valleys." What he needs is an "outlet" for the emotions that he is keeping bottled up inside. He depended on his friends, his "support group," who would listen and offer advice, including criticism at times that he didn't always want to hear. Losing them has not made him regret that dependence. If anything, the import of his remorse was that they should have depended on each other even more. In the norms of relationship that shape Dwayne's concerns, what we might call "norms of solidarity," sharing burdens with others is what friends are for.

For Dwayne, rising above his predicament is a matter of forming new bonds of solidarity and mutual dependence. He is looking for a new support group and is getting more involved in his church. As he says, "when you don't have anybody else to talk to, you can always talk to God." The idea of taking medication is simply unthinkable to him, and he doesn't regard counseling as an option. He believes there is a stigma among black men toward therapy, a view he shares. But, in his subculture, emotions *can* still be shared with friends.[15] In fact, the case most like Dwayne's is that of Ernest, forty-seven, and also African American. Loss of work due to a medical condition followed by the sudden death of his fiancée left him feeling listless, sad, and inadequate. He sought professional help and was prescribed medication, but he doesn't believe the drug is having any effect. What has made a meaningful difference, he says, are opportunities to connect with and help others, such as with the people in his mother's apartment building. These connections of affection and purpose have "saved his life."

### Emotions Reasonable and Efficient

Across participant stories, norms of relationship are intertwined with norms of feeling.[16] While the reported emotions vary, evaluation consistently involved standards of control, precision, and efficiency. For Rob the operative norms specify that after each breakup, he should have shrugged off the relationship, his emotions should have cooled, he should have moved on—quickly. He construed some of his strong feelings as overwrought, too intense: emotions must be reasonably proportional to their triggering event. He regards his bad feelings as continuing long after they should have stopped: emotions have appropriate and predefined boundaries of duration. He marks his emotions as unruly, not fully under his control: emotions are to serve instrumental purposes and self-direction. The problem is not with the particular feelings or their objects, but in himself for having experienced them.

To see the ideal of emotional efficiency in a different, performance-oriented context, we might consider the case of Rachel, whose suffering began in the tenth grade. She received a B+ grade on a calculus test and had what she calls a "mini panic attack." Long on the gifted-and-talented track in school, calculus was her first really hard class. Crying, heart racing, hands shaking, she went to the office of a favorite teacher. Unable to calm down, the school called her mother, who came and took her straight to the family doctor. That day, the doctor diagnosed Rachel with GAD and put her on an antidepressant. She took it till the end of that school year and then stopped as an act of defiance toward her mother, only resuming, of her own volition, in college.

A grade of B– on a freshman-year midterm exam got her worried. She was very upset, cried a lot, and doggedly pursued remedial help. Anticipating more hard classes to come, she went back to the doctor and was pleased that he agreed to restart the medication. Rachel divides her anxiety, which she says is only for school, "quantitatively" into adaptive and maladaptive types. The worry that keeps her studying hard is her "adaptive" anxiety, and it is crucial to her sense of purpose. "I have very high standards of myself," she says, "and I feel like I should always do the best I can, and I think that's who I am." Her worry, however, can become excessive at test time, not on par with those around her. This is her maladaptive anxiety, which she thinks must be rooted in some faulty biological mechanism because she doesn't think she is under any external pressure, including from her parents. The medication, she believes, gives her greater preemptive control of this "biological anxiety." It helps her to worry in a more optimal—"smaller quantitative"—and efficient way, still motivated to get good grades but not as prone to fatalistic concerns about getting a B or a C.

The norms of affect management used to measure experience by participants like Rachel, Rob, and others shared a broadly similar rational and instrumental focus; their goal was an emotional register precisely tuned to their control. They bring to mind the popular self-help literature, where emotional control and competence has emerged as a major theme over the past two decades.[17] One prominent example is *Emotional Intelligence*, a best-selling book by Daniel Goleman that has helped launched a veritable movement in the world of human resource management and training. The concept, as defined by Goleman, refers to a set of intrapersonal and interpersonal competencies. The intrapersonal abilities are knowing one's emotions and how to monitor them, manage them, and channel them into self-motivation and drive; delayed gratification; and an ability to embrace flexibility and adaptability. The interpersonal skills are recognizing emotions in other people and handling relationships by managing others' emotions. Goleman captures the standard for emotional expression in the simple word "appropriate." The person of high emotional intelligence is one whose "emotional life is rich, but appropriate."[18] This means, among other things, that our emotions are always in sync with our purposes, never given to extremes or outbursts or impulsiveness, ready to go into action when needed and shaken off when interfering, and rarely prone to rumination or guilt or anxiety. Appropriate emotion is emotion that can be marshalled to our instrumental ends. For many participants, just such standards of "appropriate" emotion were the terms in which their own emotions were found wanting and toward which they aspired.

### Selves Autonomous and Responsible

In participant stories, the norms of contractual relationships and efficient emotions reflect and are an expression of another social norm. Rob's emotions are "odd" to him not only because he believes they are excessive but because they defy his volition; his continued caring is not his choice. What makes some of Rachel's anxiety maladaptive is that she doesn't have control over it. With medication, she says, "I can control what I do, and I can accomplish what I want." In these stories, the turn to a relational self-sufficiency or emotional efficiency is a turn toward greater personal autonomy—greater self-direction, choice, calculability, adaptability, and control. For most people, autonomy was not merely a matter of what the philosopher Isaiah Berlin called "negative liberty," a freedom realized in terms of an individual independence from others or when no obstacles stand in the way of pursuing desires.[19] The norm also included a positive responsibility to use one's freedom to set and pursue self-chosen goals, seeking to become and remain free of dependency or other constraints, such as unruly emotions.

Possessing the requisite independence was often evaluated against another type of dependence, an insufficient freedom from the expectations or opinions of other people. Rob's anger at his friends' suggestion that he needs thicker skin points toward this norm. He is not dependent on their opinion; he determines his reality for himself. A number of people made similarly sharp comments along these lines, acknowledging there were social expectations but emphasizing that they set their own.

Gretchen, thirty-five and working two part-time jobs, measures herself against this strong norm of autonomy. Although she has been in and out of therapy at various times and is taking an antidepressant, she describes her predicament as struggling with an ill-defined and "self-diagnosed" depression and anxiety. She has trouble in unstructured social interactions with people she doesn't know well, where she feels awkward, like she is "weak" or "the most embarrassing thing" in the room. She also speaks of a more general and inchoate sense of failure. Gretchen defends her decision to "do her own thing," yet "secretly" worries that she is letting her family, other people, and even herself down. She regards this concern as a weakness. In the apparent standard at work here, you guide your life by your own choices, undeterred by—even unconcerned about—other people's expectations.

Gretchen, however, goes further, suggesting a deeper implication of the autonomy norm. She also claims that the pressure doesn't really come from others; it comes from herself. Her tacit assumption is that the self always has control over whether to acknowledge others' opinions in the first place. And, further, the self is independent even from the things it otherwise appears to value. "I'm in charge of my feelings," she says,

> So, really it's just silly feeling inferior, it's because I'm making myself feel inferior and nobody is going to make me feel inferior. So, intellectually, I know that. In practice, I have a hard time with it, and I have to tell myself that that's the case.

The presupposition appears to be that each person is (nearly) invulnerable; one cannot be directly touched or gripped or disturbed by anything outside the self, whether the judgment of others or even one's own (current) values. Being free to choose and being free of obligations or ties one does not choose means that one can only be or feel burdened or hampered if one so allows. When Gretchen feels inferior, it is because she is "making" herself feel that way; nothing else, including her own cares, could do so. Looking at the stories of other participants, including those who stressed that problems only become an illness if you let them, suggests a similar stance. In judging themselves against the autonomy norm, participants commonly presupposed some such invulnerability.[20]

Again, the self-help literature provides a good illustration of the type of self at work in the norm. In the late 1990s, *Who Moved My Cheese? An Amazing Way to Deal with Change in Your Work and in Your Life* was a runaway best seller, eventually racking up sales of more than twenty-eight million copies in forty-four languages.[21] The "cheese" of the title is a metaphor for what people desire to have in life, anything—a career, health, recognition, a family—that they believe will make them happy. The book is a parable for how successful people respond when circumstances change and interfere with or derail their own cherished dreams.

According to the book, successful people expect change and welcome it. They neither resist nor look back but anticipate change and quickly adapt to it, both in their circumstances and in themselves. They are never held back by their fears, or cling to their comfort zone, or cry over spilled milk. They know that the "biggest inhibitor to change" lies within their own mind. So, they project a future, take risks, and recognize that better things can always be ahead. They remain light on their feet, quick to learn from their mistakes, reject illusions, and adapt. In the imagery of the parable, their running shoes are always close at hand. The self, as presupposed in the book, has a sovereign power and can always stand apart, independent of its circumstances, previous commitments, and values. It is invulnerable; to live this reality, we have only to recognize it.

While a norm of autonomy was certainly a common metric of evaluation, it was not the only one. The alternative that appeared in several cases, like Dwayne and Ernest mentioned above, was a norm of solidarity. Another example is Kamesha, who evaluates aspects of her experience against both standards. A single mother of modest means, her case draws our attention to the issue of social class. The very idea of living "autonomously" and organizing life as a self-defined, goal-driven, and future-oriented project would seem to require resources, private space, and an independence from other people that only the affluent and upwardly mobile might possess. Clearly the type and range of potential choices, from matters of private consumption to occupational possibilities to lifestyles, differs sharply by class location. But the disappearance of ways of life that are handed down through tradition transforms issues of how to live into choices for everyone, not just the upper classes. What we are talking about here, such as with Rob, is a matter of aspiration, a desire to conform that can be expressed apart from and, in a sense, scaled to the available means. Kamesha is a poignant example.

Kamesha is thirty-two years old, with two sons, a teenager and a preschooler. Her life is extremely busy. She works at least eight hours a day, seven days a week, as an office assistant on weekdays and a waiter at a restaurant on the weekends. Outside work, and with some help from her mother, she cares for her sons and gets them to school. She has a boyfriend she sees on weekends and over the years has assumed daily tasks to assist nearly everyone in her life, including her siblings. According to Kamesha, most of her friends are less fortunate (and less hardworking) than she, and, in many cases, she has stepped in to essentially run aspects of their lives, from managing their finances to job hunting on their behalf. In addition, Kamesha tries to bring a little treat to her grandmother, who has Alzheimer's and is difficult and verbally abusive, every day after work.

Unsurprisingly, between work, her sons, and all the additional tasks, Kamesha reports that she is exhausted. She sees her fatigue as normal under the circumstances and her pattern of helping others as a sign of strength, of her doing the right thing. Her evaluation involves norms of solidarity. At the same time, she also appears to be measuring herself in part against the norm of autonomy and a responsibility to produce and show allegiance to a self-project. Despite her commitments and the many constraints of her circumstances, she identifies her predicament as a sense of frustration and failure for neglecting herself, her own improvement and goals. She feels disappointed in herself and, in her word, "trifling." What she should do, Kamesha says, is to put herself first and work on goals that would give her substance. (As the stories of other participants have led us to expect, her boyfriend has suggested she needs medical help.)

### Selves Optimizing and Achieving

In Kamesha's story there is a tension between the evaluative standards. Moving toward more personal autonomy seems to require less solidarity. Autonomy and solidarity point in different directions. In the predicament of Gretchen, by contrast, a resistance appeared to be at issue. She strongly affirmed living under an autonomy norm. Her stated "philosophy" was to do "whatever I want to do" and without regard to the views of anyone else. Relating to herself in this way, however, ran up against another practice of relating to herself as one with a duty to optimize her capacities in specific and socially approved ways. Part of her predicament is that she feels this duty, which she is struggling to resist, and worries that she has not made all she could of herself. The norm of autonomy involves pursuing one's own goals and taking responsibility for one's own choices. Gretchen is fine with that responsibility. It is another seemingly related norm, a norm of optimization, that is the trouble. In her view, doing her own thing should not involve such an obligation (or any unchosen obligation).[22] Since she is living the autonomy norm, she should not feel like a failure and cannot explain why she does.

For most participants, norms of autonomy and optimization went smoothly together. Among the evaluative standards expressed or implied, the responsibility to be self-steering also included a duty to take initiative, set goals, work on yourself (for yourself),[23] and achieve visible success. Resolution involved being this sort of self-optimizing person. The norm was particularly salient among the (middle-class) college students. It was not unique to them, of course. Many of the older people we interviewed evaluated themselves against the same standard. For most of these, the background assumption was that hard work at self-optimization and actual success at reaching goals were necessarily linked. Following the norm of the former would lead to the latter, and they framed their predicament in light of disappointment and confusion when this expectation was not met.

But the college students were especially likely to measure themselves in direct optimization terms. They operate in an openly competitive environment, with high expectations to excel. The norm of autonomy presupposes that choices exist in a relatively open field of possibilities, and in imagining self-optimization, students (and others) measured themselves against a "scale" of their "potential" with wide and undefined boundaries.[24] For some, even native endowment and temperament were not necessarily constraints to be recognized.[25] In this context, optimization was clearly what autonomy and self-authorship both require and produce.

Recall Rachel, who had the "mini panic attack" in tenth grade because of a less than perfect test score. For her, the imperative to "always do the best I can"—as directly reflected in grades—is deeply entwined with "who I am." Or Kristin, from the introduction, who actively pursued a prescription for Adderall. She is failing to meet a standard of making the most out of her life, which she sees reflected in her lack of a kind of celebrity status. Kyle, a college senior is the one participant who, at the time of the interview, was taking a medication without a prescription. He stresses his strong need to be productive and to accomplish things and the ease with which he can get despondent when not succeeding in school. In these moments, he agonizes over, "What the hell am I doing with my life?" Taking the medication without a prescription is OK "'cause I've done work without it" and he does not believe he is dependent on it. But it is also the right thing to do precisely because he's so much more productive and efficient as a student on the drug. Further, he says, "There has to be some form of productivity in my day. I have to feel like I accomplished a bunch of stuff, and [the medication] gives me that feeling, too." Kristin and Rachel, though with a prescription, voice a similar judgment, drawing a distinction between bad drugs, which harm you, and stimulant medications, which help you, in Rachel's words, "to make yourself better" in terms of college success.[26]

The ethical valorization of autonomy and choice has a long history in modern thought; it virtually defines what we mean by modern thought. An ideal of self-optimization and achievement has a long history too, but its current inflection reflects more recent social changes and new rights claims and institutional arrangements. Writing in the 1960s, for instance, human potential psychologists saw increasing autonomy and the liberation of subjectivity as opening the space for "self-actualization" and creating a responsibility "to express and activate all the capacities of the self," in the words of Carl Rogers. Similarly, Abraham Maslow stressed that a person's responsibility as a free agent was to make the most of themselves, realizing, in the words of a commentator, "as many of his or her potentialities as possible."[27] Indeed, they identified such initiative and maximization of one's powers and qualities, as the crucial barometers of health.

Over the decades, however, the language for expressing the norm of optimization has shifted, with other terms, like human potential, positive thinking, empowerment, "peak performance," and self-esteem, variously in use.[28] With the changes in language has come a more instrumental focus, measured not so much in terms of enlarging the self as leading to tangible and visible success. One prominent rubric that began to spread in the 1980s framed optimizing persons as "enterprising," as persons who strive, with economy and efficiency, for excellence, achievement, and self-fulfillment. "Enterprise," of course, is a business term, and it came to signify a fit between a certain corporate practice and the liberal picture of the self, a fit sometimes more recently referenced in terms of "branding."[29] For businesses, as well as organizations in other social arenas (education, medicine, family, etc.), the objective is to mobilize and utilize the "enterprising capacities of each and all, encouraging them to conduct themselves with boldness and vigor, to calculate for their own advantage, to drive themselves hard, and to accept risks in the pursuit of goals."[30] This corporate objective reflected and provided metaphors for the wider ethical and background presuppositions of the self-optimization standard.

The rubric of enterprise (or "neoliberal" as used by some) can create the mistaken impression that the yardstick of self-optimization is primarily a matter of career or economic success. While career achievement is certainly one common measure of fitness, the norm is far more encompassing than that. A modern view, increasingly predominant in our time, is that we only live once and that a good life consists in using our lifespan, according to the sociologist Hartmut Rosa, as "intentionally and comprehensively as possible."[31] This ideal is realized "in movement" and has a "logic of escalation" toward "the most comprehensive possible development of [our] talents and potentials." Given the nearly inexhaustible options available in the world and the limited number of possibilities we might realize, then the "faster one runs through the particular waypoints, episodes, or events, the more possibilities one can realize."[32] By optimizing oneself, more life projects can be completed, more experiential possibilities made accessible.

While self-optimization is measured by tangible and visible achievements, the norm is also to present oneself to oneself and to others according to optimizing qualities. These too can have a performative character—

making a good impression, looking healthy, exuding energy, appearing easy-going, flexible, resourceful, and the like.[33] Avoiding negative characterizations as passive or irresponsible, participants suggested, requires an ongoing demonstration of initiative and competence (and, again, in more than matters of career preparation or success).

Some students pushed back against this norm. Elaine, for instance, has been a high achiever, and she describes herself in high school as a "crazy perfectionist." She was active in a wide variety of activities, and her academic transcript, she says, was "perfect." Now, at twenty-three, she is nearly finished with a master's degree and works full time for a management company, where she is doing very well. In her pursuit of success, however, she has burned bridges with family and friends and does not believe that she has been concerned enough with others and their well-being. She uses the word "disconnection" to characterize her lack of feelings and being out of touch. This detachment is a source of pain and guilt for her: she feels, she says, "like I wronged them."

Elaine recognizes an imperative to "have it all together all the time" and the social approval that goes with self-optimization and achievement. She has sought to meet these norms and believes her predicament is primarily the result of being too absorbed with academic and professional success, too "consumed with accomplishments." Her mother and a friend have each suggested that she see a therapist, while she has wondered at times if she has ADHD and might need medication. But she has decided, instead, to "reprioritize." She has decided to return to her hometown and renew her relationships with her family, though this will probably mean giving up some career opportunities. She hopes that returning will anchor her and allow her to form deeper attachments. She has become self-sufficient, and she has been enterprising, but now she believes she needs to change.

### Selves Approved and Fulfilled

The imperative to have it all together, Elaine observes, creates a pressure to project a "natural, happy personality." More than any other aspect of subjectivity, we think of self-approval and fulfillment as personal and private matters. The notion of an obligation or responsibility in this aspect of our relationship to ourselves seems patently illogical.[34] And yet, we have already encountered examples of people invoking their own fulfillment and satisfaction as a kind of duty. In their impression management, from decisions about disclosure to others, to apprehensions about the performative power of words, to concerns with others overreacting or showing pity, participants stressed that one should be, and appear to others to be, happy and content with oneself as captain of one's life. Of course, desires to be more autonomous, emotionally efficient, and self-optimizing so as to overcome limitations, realize potential, move on in life, be more productive, and the like, all arise from and address dissatisfactions with self. But, for many, their evaluative judgments also included a standard of self-approval in its own right, an approval important for their own fulfillment and for the sake of appropriate (ethical) relationships with others.

In defining their predicaments, participants commonly assessed their self-approval in terms of their self-presentations. Interactions with others exhibited features of themselves that were an impediment to them and to how they conducted themselves. Their comparisons involved ideas about the requisite norms of an ideal personality, what people like and expect, the qualities that signal psychological well-being and drive.[35] Piper, for example, diagnosed with social anxiety disorder (SAD), says the ideal is to be "nice, funny, outgoing, [and] confident." Hailey, twenty-seven, a graduate student and also diagnosed with SAD, identifies the norm—what she calls "an ideal quality"—in these terms: "In our society, in every domain, it's important to be outgoing and likeable." While autonomy and independence are highly valued, isolation from others is not. Failure to convey sociability and confidence, and a positive and upbeat attitude, Piper suggests, may be taken by others as offensive or a sign that one is mean-spirited or psychologically abnormal. The norm of being is to be someone you enjoy and someone enjoyable to be around.

Positivity as a measure of self-approval also had a flip side: an injunction against self-negativity, pessimism, and burdening others with one's problems. This imperative is partly for the sake of right relations with others. Patricia, for instance, tries not to talk about her unhappiness with her friends, because "people don't want to be around negative people." Another problem with negativity, she says, is that it is self-sabotaging. At sixty-three and living alone, she is frustrated and disappointed by how life has worked out and entertains regrets about some of her decisions, including her divorce and ways in which she was not sufficiently self-optimizing. Her disappointment is complicated, however, by the conviction that happiness is "all within ourselves." Happiness and self-approval are a choice, and one we have an obligation to make. From this perspective, Patricia remarks, "I shouldn't be too frustrated. . . . I realize I have to [be] happy within myself and then share it with somebody." Self-approval is something you have to do for yourself, from yourself.

Many participants indicated that there were certain feelings that they did not believe they could express, even, in many cases, to confidants. These included experiences of discouragement and loneliness, nervousness and insecurity, concerns with physical appearance and weight, regret and self-blame. The reasons for this silence include the reasons for nondisclosure already discussed in chapter 4, such as worries about misunderstanding and overreaction. These are not easy experiences to acknowledge. An additional reason, compounding the isolation, that comes across in various and subtle ways is that such feelings violate a normative image of self-efficacy and approval, along with rules of good relations. Feeling bad about yourself is just the sort of pessimism and self-doubt that, according one participant, "pushes people away from you."

There is no better indicator of the norm of self-approval than the immense literature dedicated to helping people overcome low self-esteem. The common message in this writing is that people are scaling themselves too small, so to speak, and not appreciating their true measure. This underestimation is especially pernicious because self-esteem, according to the surgeon general's 1999 report, *Mental Health*, is a bedrock feature of mental health and individual success. And it is also, to quote the report, one of the key "ingredients of each individual's successful contribution to community and society."[36] Much personal failure, in this view, is the direct result of low self-esteem, and raising self-approval is both a personal and social obligation.[37]

The advertising copy for *The Self-Esteem Workbook* provides a good example of the ethical picture for self and others. According to the second edition, "Developing and maintaining healthy self-esteem is key for living a happy life, and with the new research [e.g., "cutting-edge information on brain plasticity"] and exercises you'll find in this updated best-selling workbook, you'll be ready to start feeling good about yourself and finally be the best that you can be." The book provides "new chapters on cultivating compassion, forgiveness, and unconditional love for yourself and others—all of which improve self-esteem." And it argues that self-esteem is an approval you can and should give to yourself:

If your self-esteem is based solely on performance—if you view yourself as someone who's worthy *only* when you're performing well or acknowledged as doing a good job—the way you feel about yourself will always depend on external factors. Your self-esteem affects everything you do, so if you feel unworthy or your confidence is shaped by others, it can be a huge problem.

> With this second edition . . . you'll learn to see yourself through loving
> eyes by realizing that you are inherently worthy, and that comparison-based
> self-criticism is not a true measure of your value.[38]

As we saw with Gretchen, the woman who feels awkward and embarrassed around unfamiliar people and believes that only she can make herself feel inferior, so here self-esteem is presented as a choice we make for ourselves quite independent of social standards or the expectations of other people.

At the same time, as the quotes from the *Workbook* imply, the self-esteem under discussion is conceptualized as fragile. Comparison and self-criticism come easy, and securing self-esteem requires special techniques to bring them to consciousness and subject them to manipulation and change. You have to learn to "start feeling good about yourself." Much of the discussion of norms in this chapter and the last has shown a similar precariousness. The standards against which participants measure themselves center on personal control and a family of social norms related to autonomy—of emotion, pure relationship, self-optimization. The exceptions I have characterized as evaluations in terms of norms of solidarity. But autonomy, as imagined in participant stories, is not simply about the self. It, together with self-approval, are also about how one should live, reciprocally, with others. The norms of being have social corollaries.

We can see the corollary of the self-approval norm illustrated in the fraught process of disclosure. Participants, recall, did not expect just anyone to honor the developing account of their suffering. Quite the opposite: "most people," according to the participant I cited, "just wouldn't understand." They had to work, often quite consciously, on the language to avoid potentially undesired interpretations from confidants and professionals and were anxious to avoid neutral auditors or disaffirmation. While interpretation was developing in dialogue (internal and) with others, there was no guarantee that the story they "choose to live in" would necessarily be affirmed or accepted by anyone else. One way to secure recognition was to appeal to external criteria or sources, such as drawing on witnesses, like a doctor's diagnosis, to validate suffering and remedial action. But there was also another way that did not depend on such criteria.

Many people appeared to take the view that they were entitled to have their story affirmed. The expectation was strongest of confidants, those in a relationship of trust. The high degree to which this expectation was rewarded suggests that confidants themselves took affirmation be to normative, that they regarded any other response as inappropriate or damaging. Beyond confidants, the expectations of narrative insularity were lower. At a minimum, others should be noncritical, fair, or "nice" (as some of the college students put it), a standard that appeared to extend to professionals. Only a small number reported neutral or disaffirming responses from professionals, but they treated those responses as objectionable. The corollary to the self-approval norm is a reciprocal obligation to be approving of the self-narrative of others. According to the *Workbook*, cultivating "compassion, forgiveness, and unconditional love" is for "yourself" and for "others."

I am not arguing in this or in previous examples for any one-to-one correlation between interview findings and such popular representations as *The Self-Esteem Workbook*. In fact, we asked participants about self-help books. A few were enthusiastic readers, but only a small number acknowledged reading them. My point is not to suggest direct influence but to highlight a congruence between participant evaluations and real cues in the cultural environment. For most people, most of the time, normative expectations are learned in the practices of everyday life not in books. Think of the pervasive language of "awesome" now used with children, which certainly embeds the view that children, as with *Workbook* readers, need to "learn to see" themselves "through loving eyes" and avoid "comparison-based self-criticism."[39] Or the relentless flattery of advertising to the same effect.[40] But for illustrating the cues, popular books are useful precisely because they typically make social norms explicit that are otherwise in the background.

## Authenticity

Although using the language of "viable," I have been seeking to show that what is at stake for participants is a vision of the good, a way to constitute themselves as an ethical subject of their own action, a way to order their life according to a normative way of being. But don't we already have a better concept than viable—a word that a few participants did use and that typically carries a distinctly ethical meaning and is often referred to as an "ethic"? The word I am referring to is "authenticity." After what I have said, at least with respect to autonomy and self-definition, might not this final section be concerned with "selves true and authentic." Much theorizing about self and identity, in philosophy and the popular therapeutic literatures, places a very strong emphasis on authenticity as an integral component of the self-created life—variously its foundation, its animating force, and its evaluative standard.[41] Popular concepts like self-actualization, self-realization, and self-fulfillment, and the paths prescribed to reach them, are all built on this foundation. The one thing that has a sacred status beyond the self itself is the injunction to "be true to oneself." Why is it missing?

In fact, one axis of authenticity has come up. In discussing concerns about medication in the last chapter, I noted participant worries about a kind of falseness or fakeness. Some, such as students taking stimulants to study better, were concerned that taking medication might be giving them an unfair advantage over others. They might be cheating or even hypocritical. Their performance might not be their own. Another concern was that medication might be blunting emotions or affecting changes in personality that were unnatural or undesired. Some of the experimenting with stopping medication was motivated by this concern. Other things being equal, for most participants, the unaltered was preferred. They were familiar and comfortable with their personality in most aspects; the desire for greater emotional control did not extend to a mere sedation or the dulling of all feeling. These features of themselves were theirs in an uncomplicated, "authentic" way. Further, in the discussion earlier in this chapter, I mentioned that with the norm of autonomy came a responsibility to make one's own path, set personal goals, ultimately decide for oneself alone. This was an issue of ownership or fidelity. One's life is one's own, and one should value it and live it in a personally fulling and generative fashion. Contractual pure relationships reflect and depend on the same consistency and fidelity to the self.

Autonomy and this ownership axis of authenticity are virtually synonymous. To the degree that autonomy was a norm of evaluation for participants, so too was this vector of authenticity. The new rule of choice requires self-ownership. It is hard to see how they could be separated. Given our institutions and practices and the decline of more traditional ways of life, there is a virtually inescapable sense in which we are bound to imagine and enact our lives, past and future, as matters of conscious choice. According to Anthony Giddens, "we have no choice but to choose how to be and how to act," or as another British sociologist, Nikolas Rose, puts it: "The modern liberal self is 'obligated to be free,' to construe all aspects of its life as the outcome of choices made among a number of options." Choices that, in turn, as Rose adds, "reflect back upon the individual who has made them."[42]

But there is another axis of authenticity, based on a quite different meaning of "true." By this second axis, I refer to an ideal that is often traced to the Romantic movement of the late eighteenth century (and later phenomenology) and its valorization of imagination, spontaneity, emotion, inwardness, creativity, nature, and the like.[43] This ideal is based on a sharp opposition. On the one hand, there is a deep, "true self" that lies within each individual. This innermost self is the wellspring of our desires, moral feelings, dispositions, and other attributes, which are distinctively our own and constitute for us critical guides to the truth of how we ought to live. By turning inward, we seek a self-transcendence, finding our place in something beyond ourselves and greater than ourselves—nature, the cosmos, the order of being, and the like.[44] On the other hand, there is the social order. This outer world, dominated by "the contrivances and conventions, the masks and hypocrisies," of society, is the realm of the distorted, false self.[45] It is mere surface and artifice, and it stands between us and our true self. Breaking through this barrier means turning inward, getting in touch with our deepest feelings, and then giving aesthetic expression to the truth and originality that lies hidden there. The inward turn involves intensive introspection and candid self-appraisal, seeking, in the words of the philosopher Charles Guignon, to "achieve genuine self-knowledge." Then, the ideal specifies that we must live by this knowledge, expressing and demonstrating in all our actions our true self—"to actually *be*," Guignon writes, "what we *are* in our ways of being present in our relationships, careers, and practical activities."[46]

With the social norms discussed above, I could point to only a small number of counterexamples. With authenticity, I could find but a few genuine examples. The reason, I believe, is because authenticity in the phenomenological sense presupposes a depth psychology and a truth beyond the self. But through the lens of the neurobiological imaginary, as I'll suggest further in the next chapter, addressing everyday suffering and moving toward viable selfhood does not. Most participants did not, for instance, take issue with themselves for engaging too little in introspection, or cutting self-searching short, or missing their unique way of being human. They made no indication that they should have followed a deep truth that was inside them, a truth that if they had but known it better might have prevented their predicament or, at least, might now guide their actions for the future. They did not voice a desire to find their place in the great scheme of things, or a yearning for a lost wholeness or unity or lay stress on its restoration in their lives. They did not communicate a hope or credit a medication with helping them raise to expression unconscious processes or prioritize their creative powers. Very few set their own experience against a false and shallow social order or faulted themselves for being too caught up in its deceptions and superficiality. Very few conveyed a hope in insight psychotherapy—the modality in its many expressions that plumbs the inner life—for much of anything.

Authenticity is missing because the path to viable selfhood does not follow such an inner-directed route. Being true to one's self can only make sense in light of a broader horizon of meaning in which the self is located.[47] The inward turn is contentless if our stories are merely self-referential.[48]

## Conclusion

In making sense of their predicament and the direction that moving forward should take, most participants, both explicitly and implicitly, measured themselves against a package of social norms—of relationship, emotion, or self—that share a strong family resemblance to the liberal ideal of autonomous being. Achieving viable selfhood meant being in personal control, independent, and with the power, to quote one participant, to "dictate where you are." In an older ethic of a self-created life, authenticity as inner truth would have stood out for answers to existential challenges. But not here. The destination is different and so is the means to get there.

Recall Rob, with whose story I began the chapter. He asserts a strong sense of autonomy, and his goal is to bring his experience more into conformity to this social norm. Society is not what is bringing him down, but something within himself yet not *of* himself. On his account, only things he chooses come from him (that is what self-sovereignty means), and since he does not choose to feel about relationships the way he does, some genetic condition must be at work. Recall Gretchen, quoted several times, who takes the same stance. She refers to her feeling inferior about not striving for high achievement as "just silly," yet her suffering is sharply painful and persistent and barely touched by a high dose of medication. She says at one point that even on her antidepressant "I wake up and the first thought is, 'Christ, my life sucks!'" She did not want to go on the medication again. However, she now thinks that "medication is important because there are physical things that I don't understand that probably occur." The "physical things" come in, if I follow Gretchen's implicit reasoning, because she does not, in fact, *choose* to make herself feel inferior. That feeling is simply unaccountable in autonomy terms. Something else, something physical and "not self," must be at work.

In this imagining of viable selfhood, there appears to be little room for either a traditional psychological or social explanation, alternatives that might account for predicaments without attributing norm violations to choices and allowing for the incorporation of troubling experience within the self-narrative. Of course, there are imaginaries—psychodynamic, hermeneutic, phenomenological, religious—that take a different route and for which the incorporation of past experience and the importance of social context would be key. The exceptions noted in this chapter, like Dwayne and Ernest, Kamesha and Elaine, are drawing from some such (perhaps religious) imaginary.

In this imagining of viable selfhood, in which the predicament consists in a constraint on volition, the neurobiological offers a powerful interpretive possibility. The source of trouble can be split off from the self. It can be detached from the participant's history, concerns, and context and located elsewhere—in a faulty biological mechanism, or, as one participant I quoted in chapter 4 said of the physical cause, in "too little or too much or whatever it is that makes you have these issues." Splitting off permits a new image of the person living undisturbed in his or her social conditions, being what it is right to be. Splitting off opens up a new way to imagine the effects of medication, as confirming the faulty mechanism and providing hope for its overcoming. With the neurobiological imaginary, ends and means come together.

# SIX

# After Psychology

The crisis came in the year after college. Hailey was suddenly on her own and in a whole new environment, faced with day-to-day responsibilities and obligations she had never had to handle alone. Everywhere she turned, she felt anxious and self-conscious. Work was especially difficult. She got nervous every time she had to make a phone call. She had trouble meeting new people, chatting with coworkers in the hallways, and going out with them after work. She felt out of place and worried that others believed that she was acting strange. She was on edge and not sure if she could continue.

Although she had never had such an intense experience, Hailey was no stranger to "issues with anxiety." Her childhood had been happy, with lots of friends, but she was not one of the popular kids and when in groups often worried that she would say the wrong thing or look foolish. Her nervousness around others may have kept her from projecting a cheery disposition, because she remembers her friends encouraging her to smile more and so not appear rude. At school, she avoided talking in class, and when she had to give presentations, her hands would shake, and she couldn't think clearly. Her teachers regarded her as shy, according to her mother, who encouraged Hailey to be more outgoing.

Hailey didn't like being thought shy. It felt like a criticism, she says, "like something was wrong or I was abnormal" and needed help. As she got older, however, Hailey came to share the assessment of shyness as negative and a limitation. "In our society," she remarks, "it's really important to be able to talk to people and to be able to make connections." When she was falling short of this norm, "I definitely felt really bad about myself," and she came to feel that she was deficient in this important area of life.

As the postcollege crisis grew, Hailey sought medical help. With a degree in psychology, she already knew a little about social anxiety disorder, that it was considered a common problem, and that there were "biological processes," involved, as she puts it, "in terms of dopamine and serotonin." In fact, before seeing the psychiatrist, she had already appropriated a SAD diagnosis and medication. "I knew that this was my diagnosis," she says, and a prescription was "what I was looking for."

The psychiatrist diagnosed Hailey with SAD and wrote a prescription for an antidepressant. Hailey noticed less nervousness right away, but was still feeling out of sorts and decided to see a psychologist. The therapy didn't last long. She reports:

> I don't know, I wasn't really into it. I didn't see it helping as much as the medication, especially once I started taking the medication and I realized it was helping me I just didn't see the need. I think looking back, I'm glad that I did go to the psychologist for a couple sessions because I think the cognitive behavioral tricks that he taught me were pretty useful in some situations. I don't really use them anymore.

As she continued on the medication, she began to feel more confident in social situations, like a "different person." She was able to relax and grew less worried about what other people were thinking of her.

While Hailey used to think she was shy by temperament, her prior knowledge "of how the neurotransmitters work" and her experience with the medication brought an impression point. The shyness, she now believes, was caused by "a biological genetic thing." She knows that SAD is a mental disorder, but she regards "disorder" as a flexible and relative term and she demarcates a boundary between her appropriated use and what "according to researchers" SAD might mean. What her experience means to her is that it confirms that something neurobiological was amiss, a "blockage" or "screen" in her brain. And it confirms, she says, that this was "definitely something I needed medication for."

Now twenty-seven and a graduate student in psychology, Hailey sees the removal of this screen of anxiety as giving her new options in life. She doesn't "feel limited anymore" and, on track to have a career, no longer worries about what she is going to do with her life. Now, she says, "I feel like I can do anything." Her only regret is that she wasn't diagnosed and given the medication earlier. She could have been more optimizing. For instance, while she was very happy with her college experience, on medication she might have run for class president.

Although Hailey spoke of feeling like a "different person," she stresses that medication has not, in fact, changed her in any important way. She does act differently. Previously reticent in class, she is always talking now and even blurts out comments without thinking. But this "extreme," she says, doesn't bother her and doesn't represent a real change. She is not really a different person, just better able to express herself, and with the same motivations she had before the medication: "I'm able to do things I couldn't do before but I wanted to do. It's not like I suddenly wanted to do them. I always wanted to do them I just couldn't do them." For Hailey, her goals and purposes are her own, now finally released from the physical blockage of the anxiety that was interfering.

---

In the last chapter, I presented the social norms and ideals of being, whether directly articulated or in the background, that participants drew upon to access their predicament and define resolution. These norms centered on autonomy, and, for most people, ameliorating or rising above their predicament meant bringing their experience into line with a more independent, self-defining picture of themselves. I also argued that most did not use authenticity as an evaluative standard, at least in the common phenomenological meaning of authenticity—as getting in touch with and expressing a deeper, inner self from which one can draw resources of insight and truth.[1] In their efforts to move toward or restore a viable self, participants made little reference to their inner life or conflict with society. Even Hailey, in graduate school for psychology, doesn't see much need for any sort of counseling. And when she does go, her few sessions are geared to instrumental problem-solving and optimized functioning. She got help to "rationalize" her anxiety, she says, by asking what sort of cognitive errors she was making, and then trying to correct those with various "strategies" and "tricks." If she thought her anxiety might provide a clue to some inner truth, or have some function in her life, or be a sign of some social unsettledness, she makes no mention of it.

Rather than exploring memory, evaluative outlook, or social situatedness, for participants like Hailey, the process of establishing or restoring "viability" involved something quite different. This process might best be called "fixing." Fix is one of those frequent participant terms that conveyed an instrumental, almost mechanistic approach to self and everyday suffering. A few of the psychologically minded contrasted the work of therapy with a dismissive assessment of medication as a "quick fix." Far more commonly, participants used the term "fix" positively, referencing brain problems or the action of medications to indicate that a problem could be resolved. "That was just a messed-up thing in my brain," said one, "that needed to be fixed." Like Hailey, they regarded their suffering in terms of mechanistic impediments, tangible and objective, to be corrected or repaired or removed, not as experiences to explore, or elaborate, or interrogate their self-understanding about, or even to heal from.[2]

Further, with some qualifications, people like Hailey attributed specific desired results directly to the physiological action of the medication, even in cases, such as we saw with Jenna in chapter 4, where they emphasized that the effects of the drug were subtle and hard to discern and could require professional assistance to interpret. In this view, drugs operate outside of narrative, bypassing all social influence and requiring neither interpretive work nor social support. Acting on the central nervous system, they directly shape thought, emotion, and conduct. For Hailey, what she is offering is not a self-interpretation but a description of a newly discovered truth. But if we reflect on what participants like Hailey actually say about how the fixing is envisioned or believed to be accomplished, we do not find an explanation of neurobiological processes. We find a narrative of self at work to reduce tension and reach toward conformity with particular social norms.

In this chapter, I explore another dimension of the background picture that makes the neurobiological imaginary compelling to think with. In the narrative trajectory of medicalizing accounts, the neurobiological conception of causation is joined to various changes attributed to medication.[3] In order to see these changes, I begin by contrasting accounts like Hailey's with the stories of those who took medication but with a psychologizing perspective. The difference highlights how the neurobiological facilitates the quest for viable selfhood by opening up an interpretative possibility that would not otherwise be available. The imaginary offers a way to envision oneself with unencumbered volition. By recourse to a technical and objective intervention, one can disengage the normative deficiency and begin to realize authorship of one's own life.

The technical intervention is medication, its effects interpreted as splitting off aspects of experience from the self. If the regime of the self as I have described it captures important social norms, then achieving viability in its terms seems straightforwardly desirable. In many contexts, this is how we are expected to live, and a social cost may be paid if we fail to do so. By contrast, the value of imagining medication as annulling aspects of one's history and context is less immediately obvious. Taking a medication and benefitting from it, as we'll see again in this chapter, does not require such an interpretation. Something else makes it appealing, and I next turn to some reflections on features of the social organization and pace of how we live now. Living in contemporary society, in our "liquid times," I argue, fosters a light, depthless way of being, a way of being that comports with the regime of the self and the means of the neurobiological imaginary to envision it. In fact, the imaginary represents a kind of theodicy that reconciles normative order and suffering with the liberal vision of self-designing.

## Medication Effects

In talking about changes brought about by taking medication, a number of participants referenced the power of magic. Some, such as Lisa in chapter 1, worry that they might be heard as attributing too much efficacy to the drug and stress that other changes have also contributed to their improvement. The medication helps, but it does not work like magic. In a similar way, Eric in chapter 2, though he seems to wish that his medication had the magical power to make "everything awesome," says that it just helps him cope. Others, however, referenced magic in an opposite context, of too little rather than too much effect. They didn't know what the medication was supposed to be doing and could not discern any specific changes. Was something magical supposed to happen, they asked, that they were missing? Gretchen, one of the people who had responded to a drug advertisement, wonders, "Am I supposed to feel like skipping in a meadow with daisies and singing a song with like unicorns and rainbows? Is that how I'm supposed to feel, because I'm not." Her doctor has been adjusting the dose of her antidepressant and asking Gretchen to judge whether more or less is necessary. But, she asks, "how the hell should I know?"

All these references to magic highlight the role of patients as the evaluators of the effects of medication in their lives.[4] Whether the drug is "working" is based on their self-report and is commonly accompanied by the assumption that specific drugs alter specific emotional, behavioral, or cognitive states or processes that the patient should be able to accurately identify. This assumption is rooted in the specific disease model used in the DSM and lies behind the classification of psychotropic compounds with specific types of disorder: antipsychotics, antidepressants, anxiolytics (antianxiety), "ADHD medications," antimanics, and so on. Specific types of disorder have a corresponding medication treatment. The same notion also informs the way medications are marketed and discussed in doctor-patient interactions.

But scientists have long known that the physiological effects of psychotropic drugs are largely nonspecific, with wide variations between individuals. Consider the class of drugs called "antidepressants." Epidemiological studies show that these drugs are commonly being prescribed for such problems as anxiety, chronic pain, neuropathy, fatigue, and sleep disorders.[5] In other words, many things in addition to depression.[6] There are also pronounced differences between people in their response to these medications. Among interview participants, some responded to one of the available antidepressants but not to others. Some tried several brands before finding one they thought helped (and did not make them sick or dizzy or tired). Some found the antidepressant to lose effectiveness after a period of time. Some found one ineffective but felt better when taking two in conjunction or one antidepressant and one other type of medication. Some could not find any antidepressant that helped. Virtually every person who was not new to medication had changed at least once.

Further, talk of a medication working can give the erroneous impression that the meanings that a drug has for a patient are simply given or flow from the biochemical properties of the drug itself. There is a large body of research showing otherwise. Assessment, for instance, requires patients to establish a cause-and-effect relationship between the drug they have ingested and any physical sensations or other changes they notice about themselves. As we have long known, this relationship is easily and commonly misjudged.[7] In randomized controlled drug trials, for instance, people in placebo control groups often attribute changes, relief of symptoms, and undesirable "side effects" to the chemically inert substances they think are medications. In the case of antidepressants, the average placebo response rates in trials since 1991 has consistently been in a range between 35 and 40 percent.[8]

The placebo response is not unique to psychoactive medications, of course. It is found throughout medicine. It arises from the positive meanings and symbolic values that interventions have for people. Clinical studies indicate a number of interacting contextual factors are relevant, from individual and professional beliefs and expectations, to the clinician-patient relationship, to aspects of the surrounding social environment and treatment setting.[9] A number of these nonpharmacologic influences have already come up in the participant stories I have recounted. Among other things, people commented on the importance of friends and family members for exchanging ideas about drugs and their expected results; family, racial group, and confidant attitudes toward drugs and taking them; the perceived response of others once the medication regimen began; confidence in the prescribing professional, and sometimes psychotherapist, along with his or her affirmations and explanations; the meaning of the diagnosis as conferring reality on suffering; and past experience with medication.

Behind these specific influences is the much broader set of hopes and convictions that circulate in society. In the case of medications, a great deal of power has come to be ascribed to particular classes of drugs and specific brand names. I gave examples in chapter 2 of the intense popular hype that surrounded Miltown, Librium, and Valium in the tranquilizer era and the celebrity status of Prozac (and the SSRI class), Ritalin, and Adderall in the more contemporary period.[10] Well in advance of ever taking them, potential users have encountered messages about the transformative effects of these drugs and known or heard about people finding relief using them. One element of the neurobiological imaginary is just this belief in the power of medications, acting by themselves, to fix problems—a belief that even those who had never taken a medication shared and which remains strong despite growing popular criticism of side effects and even suggestions in some quarters that the SSRIs are nothing but placebos.[11] The social reputations of medications condition the meanings that people (and this, of course, includes clinicians) attribute to them, both as an object and, if taken, as an experience.

Appraising the changes brought about by taking a medication, then, including the physiological effects of the medication on the brain, involves a cumulative learning process. The meaning is not predetermined, lodged in the medication itself; it is imputed by users and shaped and reshaped in the course of their interactions with the drug, with those persons and sources of information with which they are in contact, and other changes they are experiencing.[12] As we have seen, even seemingly very similar experiences, such as the blunting of feelings reported by people on antidepressants, can be read either positively or negatively. Similar to diagnostic categories, medications have the characteristics of what the anthropologist Victor Turner called "multivocal symbols." They can represent "many things at the same time," with referents that "are not all of the same logical order but are drawn from many domains of social experience and ethical evaluation."[13] In order to understand effects attributed to medications, this multivocality has to be recognized.

Among the people I interviewed, the meaning of the drug and the evaluation of its effects was tightly coupled with the how they interpreted their predicament and its possible resolution. The definitions of predicament and efficacy were interwoven and unfolded together. Among those on a medication making sense of experience followed two general story lines, depending on whether they explained their predicament in psychologizing or in medicalizing terms. The former, in which medication is emplotted without biology, is a story of emotional stabilization. The latter, which emplots the medication with a biological cause, is a story about the removal of an external constraint. In tying their predicament to the brain, this latter group ascribed (or expected) much greater self-transformative efficacy to (or from) the medication. And, generally, in contrast to the psychologizing, expected that their use might be long-term.

I need to mention from the outset that the majority of people, even the most enthusiastic, expressed some disappointment with the drug they took. Recall Rob from chapter 5, who felt the medication was having a diminishing effect, or Kristin from the introduction. When Kristin, impressed by the vivacity of a classmate on Adderall, concedes that the medication is "not like a wonder drug," she reveals her frustrated expectation that something exceptional would happen. Many others spoke of their letdown in expectations: it "wasn't like happy pills," it didn't "take away all of the anxiety." Medication helps, they affirmed, but it does not completely resolve their predicament or end emotional pain. But even if it did not actually deliver desired feats of self-transformation, what it did do narratively (as a symbolic object) is support and affirm the causal interpretation and the sense-making that interpretation offers.[14]

### Medication without Biology: "I Just Felt More Stable; Not as Emotional"

About one-third of those on a medication took a psychologizing view of their experience, and in their stories their own will to overcome adversity is central. A few people attributed no positive effects to the medication (antidepressant, stimulant, or anxiolytic) and suggested that change is something they simply have to do for themselves. The others described the medication they were taking, either an antidepressant or anxiolytic, as quite helpful, although they often reported that they found it difficult to identify which changes might be attributed to the medication and which to other things. Though not a matter of sharp distinctions, participants generally described the medication effect in terms of helping to ameliorate a crisis by bringing increased emotional control, renewed energy (often because of better sleep), more mental focus, and greater enjoyment of and interest in everyday activities. Some also expressed a seemingly separate idea that the medication made them feel good. These effects, combined with other positive changes and relational support, brought a balancing or leveling of emotional experience among other improvements.

In these stories, medication had a role, but it was circumscribed by the larger narrative of individual effort. Linda, to give one example, described her predicament in terms of an "emotional midlife crisis." She was feeling overwhelmed, and her emotions—both negative and positive feelings—interfered with her ability to figure out what was going on. With the emotional regulation she attributes to the medication, she can now reason better, allowing her to identify and separate ("sort") the various causes of her distress and deal with them, as she puts it, "rationally." In this way, participants credit the medication with helping to return or restore them to their emotionally more balanced and in control "former self."

While emphasizing emotional control and stabilization, some people also commented on other features of the medication experience. Some of this experience is judged negatively, including some personality changes and continued bad feelings. Some is judged positively. Linda believes that the drug has changed her, making her "more lackadaisical" and of the attitude that "whatever happens, happens." She likes this change and thinks it gives her more choice. She illustrates with a story about a job she took, prior to medication, at an upscale clothing store. She wanted to learn to be more sociable, and the retail environment would force her to interact with people. But once on the antidepressant, making such an effort no longer seemed necessary. She decided she was just not the "social type of person" and "was not any the worse for it." She quit the job. In this way, she sees the medication effect as bringing not only a restoration of emotional stability but also a change of self that augments her autonomy with a different sense of the goods in which she should invest herself.

### Medication with Biology: "I Actually Do What I Want"

Those with a psychologizing perspective attributed the primary effect of medication to restoring a former, and more viable, self of emotional stability and rational action. Those with a medicalizing perspective told a different story of medication effect. This story came in several different versions, but its end point or expected end point was the emergence of a more ideal (ideally viable) self. In these stories, personal effort is not central. Rather, participants attributed change to the drug's removal of a constraint, a problem in the brain that had to be dealt with externally and mechanistically. Consequently, the medication itself is an important character in their narratives.

Although these participants invoked a neurobiological explanation, theirs is not a story about the physical workings of the brain or about science or even about an illness. None, as we have seen, can say much about the "something biological" awry in their brain or how the medication works on them physiologically, and none really try. To a degree this disinterest might be expected, especially as no one seemed to regard the brain problem as a serious threat to their physical health.[15] As one person expressed the common view: "something is a little off." But to this lack of any specific knowledge, they also conveyed a lack of curiosity or interest. Eric, for instance, the teacher we met in chapter 2 who had a "meltdown" under the pressure of teachers and parents at his demanding school, does not understand the neurochemistry and does not care to know more. No doubt Eric and other participants like him believe there is a science out there that can explain the truth about the brain, about abnormal emotions, thoughts, and behavior, and about how medications produce their effects. But in their accounts, the nature of this science appeared to matter little. Their references to neurobiology were not so much about the actual workings of their brains as about limitations on their volition.

As many of these people noted, what having a "chemical imbalance" means, to quote Jon again from chapter 3, is that "it's not my fault." And it is not his fault because, to quote another participant, Sarah, the suffering is "something that's brought on by something other than myself," something, she continues, "in your brain like different hormone levels or I don't know, neurons firing at the wrong time." In this understanding, "myself" is not my brain or my body (tellingly, Sarah speaks of "myself" but "your brain") but the conscious seat of individuality and control. Associating limitations with the brain was a way to disown them, to move them away from the self to a mechanism (brain) outside the self.

Hailey illustrates how this separation is achieved and how a more ideal "hidden self" can be revealed in the process.[16] For those with a psychologizing perspective, medication use could represent a kind of falseness. On the face of it, Hailey's use of a drug to achieve an ideal would appear to raise a similar issue, the question of whether the medication is working to enhance (change) her personality by making her more outgoing. She is quite aware of this implication and goes out of her way to stress that the drug is not changing her personality or her goals. Affirming a continuity with her premedication experience, however, is inconsistent with her account of her shyness and self-consciousness as a child and her struggles since college. These experiences suggested to Hailey that shyness and self-consciousness are persisting aspects of her personality. It was this conclusion that led her to think she was abnormal and to feel bad about herself—she seemed unrepairable, self-limited, and unable to meet the social norms of extroversion and positivity. She wanted to be free of herself because she was failing to be the person who could live according to the norms. Her desire to conform was not enough.

For Hailey the issue of her personality is where the "biological genetic thing" enters her story. With the neurobiology, she redefines her past experience in such a way that it allows her to identify the person she actually *is* with the ideal and future-oriented picture of herself as norm-conforming and free of limits. Her redefinition turns on converting her previous inhibition, anxiousness, and self-consciousness into an object that has no bearing on who she is. By this transposition, a story of trouble with her settled personality is replaced with a story of a neural blockage. Envisioned as something outside the self and the mind, the blockage can be mechanistically removed, and this is how Hailey explains the drug effect. In the new story, the drug does not enhance a shy person because she was never truly shy in the first place.

In Hailey's retelling, despite all of her actual success, she has been in the grip of a profound misunderstanding. She treats the brain mechanism as an independent actor that has selectively shaped aspects of her self-experience outside of her awareness, her intentions, her evaluations, her relationships, or the life she has led. It is as though an important part of her premedication everyday life belonged to someone else. This unchosen experience and only this experience—the valued aspects of herself are hers—is caused by the faulty biology. Isolating it away from the self establishes her viability and autonomy, and this interpretation both guides and is confirmed by the medication effect. By removing the "blockage," a metaphor that captures her personal experience of feeling hindered and limited in social situations, the drug helps her to actually perform more to her desire—garrulous, assertive, networked—and be the outgoing "people person" who was there all along. The drug helps to reveal her hidden viable self, a revealing, it seems safe to say, she would never have envisioned with her CBT or an insight-oriented psychotherapy.[17]

Hailey's story is one form of medicalizing narrative in which medication is credited with helping produce the self that it is good to be. Of course, a medicalizing explanation, by reframing experience in purely somatic terms, already symbolically removes that experience from any evaluative or moral framework and so often generates a more favorable self-interpretation. But in stories like Hailey's, another element is at work. By a kind of subtraction of experience, the establishing or restoring of the good self is attributed to the action of the drug itself. We saw examples of this with Lisa and Eric, from chapters 1 and 2, respectively. Both interpret the drug as fixing the neurotransmitter imbalance that caused their immature or unethical actions and allowing them to disavow things they had cared about and responded to that they now regard as inappropriate. Those around them, they also emphasize, see the change and regard them as better people.

Some of the clearest examples of this interpretation came from college students taking stimulant medications. They spoke of the medication as removing an obstacle, some feature of themselves or their experience, that inhibited their motivation to do what they were supposed to do or to achieve their potential. Both points were often made with reference to the notion of a "level playing field." The point of the metaphor, which is commonly referenced in pharmaceutical advertising of the stimulants, is precisely to counter the notion of enhancement.[18] The effect of the medication is simply to correct the broken mechanism. It does not confer an advantage; it brings the "real me out to light," as one participant expressed the idea.

In participant stories, however, the level-field metaphor seemed to mean something different. What they attributed to the medication was not so much that it allowed them to play the game by the same rules as others but that it gave them the ability to play their own game better. The determination that the field was un-level, for instance, was typically made on the basis of struggles in the context of new and very specific challenges—tougher versions of classes; paying attention or doing assignments in classes they found of little interest; struggling to get good grades without all the extra credit opportunities provided in high school—or on the perception that other students had an easier time doing their work. The effect sought or credited to the medication was not the revealing of a hidden self (they had been able to comply with norms of high achievement in earlier and different circumstances) or a former self either (they noted that the circumstances in which they were experiencing failure were in fact new and different). Rather, the effect was to remove limitations and so produce a more viable self, better able to meet the higher standards.

From both the psychologizing and the medicalizing perspectives, participants ascribed meanings to medication that corresponded with their perspective and reaffirmed its import in their self-narrative. That feedback loop was not all that was taking place, for the practice of self-interpretation was itself affected by the experience of medication. People had to interpret physiological changes they were feeling or make sense of the fact that they were not feeling any such effects at all. They had to interpret which changes were positive and desired and which were negative and therefore side effects. They had to interpret which effects they could own as theirs and which might be false and unnatural. They had to interpret which effects came from the medication and which came from other help or other changes in their life. So, the taking of the medication itself created new experiences and induced unexpected changes—as in the story of Linda, who believed the drug made her "more lackadaisical"—that called for interpretation and evaluation.

That said, the causal framework and medication effect were clearly in a feedback relationship. In the psychologizing perspective, individual effort is central, and the drug is conceptualized as aiding this effort and (in most cases) nothing more. In the medicalizing perspective, by contrast, the medication effect is central. This positioning corroborates the symbolic separation of suffering from self and from specific social conditions, as well as the decisive role played by an external mechanism beyond the reach of the individual to change. It confirms the anomalousness of the predicament, that it was arising from a force outside one's circumstances, understanding, or control. It confirms that the problem is "material," and therefore real: sufferers are the "host" of a somatic malfunction and are neither its source nor its solution. It affirms that the drug is the primary technology for fitting oneself to the relevant norms of being.

## An End of History

Hailey's story shows a crucial dimension of what makes a neurobiological account compelling. It opens up a new interpretive possibility, a way to make sense of suffering, settle the question of what has gone wrong, and imagine a resolution. What Hailey found was a way to explain her predicament and envision a future that went beyond gaining more control to *being* more normal in just the way that American society requires. Her neurobiological account, sanctioned in her view by scientific authority and confirmed by the reduction in inhibition brought about by the medication, allows her to detach her limitations from herself, and so be the person who is no longer different or held back.[19] By stripping away certain features of her history and context, the tension between ideal and insufficiency is seemingly resolved.

Hailey's story is a particularly stark example of how neurobiological terminology and technology enable the envisioning of a self free of the past and past encumberments. It contrasts with psychologizing stories, where medication is envisioned in terms of restoring rather than revealing the self. What distinguished most of the "psychologizing" (whether taking a medication or not) was an emphasis on individual and (mostly) unaided effort. This approach did not require any "presentist" orientation to self-understanding or a symbolic detachment of anomalous experience.[20] For a few at least, and broadly consistent with nonneurobiological perspectives—such as the psychoanalytic, psychosocial, hermeneutic, and those within specific religious traditions—imagining suffering without temporality or social context makes no sense. According to these participants, the significance of things is as revealed in their limitations, vulnerabilities, and frustrations, as it is in their capacities, strengths, and successes. Reflection on beliefs, emotions, and desires is important because blindness toward or ignorance of their evaluations and motivations is debilitating. There is a truth about the self, a faithfulness to reality to which they are answerable but about which they might be in error or confusion. Self-reflection was challenging because there was the risk that they might get things wrong, that they might be deceived, that they might not like what they found.

Monique, for instance, age fifty-six, and unemployed at the time of the interview, was diagnosed with depression during a very difficult period in her life, when her marriage was falling apart and she suffered a miscarriage. She briefly took medication prescribed by her doctor but soon stopped. Instead, she sought counseling. There, she says, she began to "work through" her very difficult relationship with her mother, whom she characterized as abusive to her as a child. She kept this a secret, however, until she says, "I talked about it. I told the secrets, I brought it out, I examined it, I laid it on the table, I no longer held it in. And so, yes, the [CBT therapy] for me was very important because I had to do some inward analysis. I got to look at what was real." Once the source, the "real issue," was identified, brought out, and expressed, she could work to understand its effects and cope better with them.

Stories like Monique's, however, even among the "psychologizing," were uncommon. Only some stressed unique biographical experience, or features of the normative environment, or the need to dialogically explore the meaning of their feelings or actions. Only some, in other words, saw a deeper engagement as the response to a predicament. Others seemed to imagine their suffering in more mechanistic terms, like the medicalizing. It is hard to say more definitively. Most alluded to the working of some non-rational factors (sometimes using a language of "disability" or diagnosis) in their suffering without identifying what those factors were, while also insisting that their psychic significance, if any, was obvious to them. Many had a fairly fatalistic view of their own potential efforts to change, with a distinct openness to turn to medication. Some of the psychologizing, then, appeared to imagine themselves in this regard neurobiologically, even without explicit reference to the brain or a specific mechanistic solution.

A self free of the past and a self-mastery through disengagement from the unchosen is what the neurobiological imaginary holds out. Via this route, Hailey is now authoring the story in which she chooses to live, identifying herself with her will and her ability to make a choice and build a world at least privately validated.[21] Ironically, in light of the promise of freedom, self-authorship in these terms embeds Hailey more firmly in a cultural framework. Her self-forming activity is now oriented even more toward compliance with the dominant norms. In fact, another reason why she need not engage in a qualitative elaboration of self is because the good for her is already prespecified. Hailey is shy and self-conscious and had, at least for a time, been somewhat overwhelmed by her new adult responsibilities. These were her symptoms, and in the biomedical framework, they are not a sign or indicator of anything else, of anything psychologically enigmatic or complex or whose meaning itself must be explored and uncovered. They *are* the problem. Hailey just needs to be more outgoing. All she needs to know is right there, on the surface.

In stories like Hailey's, a little pill has a big symbolic effect, suggesting a strong affinity between the imagined, objectivist means and the desired self-mastery. But why might that be so? What experience of ourselves must be operative for this close relation to make sense, to seem fitting?

## Selves in Liquid Times

In the last chapter, I referred to the dominant norms of self as a "regime," and I meant that not only in the sense that they represent a consequential cultural scheme for governing one's relation to oneself but also in the sense that they fit together into a kind of value regimen or package to which many participants shared a commitment. In describing evidence for this system, I mentioned some ways in which broader changes in society were effectively throwing us back on ourselves, undermining stable patterns and life-ordering criteria, and forcing us to enact more of ourselves in terms of conscious choices. Such change, social and technological, has had profound consequences for our sense of ourselves, both contributing to our predicaments and, especially to the point here, influencing our responses to them.

Rapid social change has been the hallmark of modernity. Regarding the nineteenth-century "bourgeois epoch" in which they lived, for instance, Marx and Engels identified its defining feature as the "constant revolutionising of production, uninterrupted disturbance of all social conditions, everlasting uncertainty and agitation." By this dynamic, they famously argued, "All fixed, fast-frozen relations, with their train of ancient and venerable prejudices and opinions, are swept away. . . . All that is solid melts into air."[22] Like other social theorists of earlier eras, Marx and Engels celebrated the breakup of traditional social relations and practices not because breakup itself was a good but because these "ossified" structures and old hierarchies stood in the way of building better, more rational institutions. And for a time, ordinary people found niches in the new modern institutional and class structures that, while perhaps as "stiff and indomitable as ever," to quote Zygmunt Bauman, provided some stable orientation points and meaningful structures by which they could find direction and guidance.[23]

It is only more recently, especially in the last half of the twentieth century, that change has become a kind of end in itself. We now find ourselves living in "liquid times," to use Bauman's apt metaphor for a social world that is in constant motion. Fundamental shifts in production and consumption, the restructuring of work and family, and the spread of new media and electronic technologies are among the key changes that social theorists have identified as producing a more fluid, plural, and unpredictable social environment and dramatically changing the lifeworld in which we experience ourselves. A turning point came in the 1960s and 1970s. During this era, the French sociologist Alain Ehrenberg observes, "In all areas—be they working life, family, or school—the world was changing its rules. . . . Each individual had to be up to the task of constantly adapting to a changing world that was losing its stable shape, becoming temporary, consisting of ebb and flow." These institutional transformations required new things of people and from them; they "made it seem as if each person, even the humblest and lowest of the lot, had to take on the job of *choosing* and *deciding* everything." It was only in the context of this new fluidity, he argues, that the strong value of self-sovereignty—that each of us is the sole owner of himself or herself—became a broad-based sociological phenomenon.[24]

There is much evidence that Americans, responding to social change, were shifting the way they anchored themselves after midcentury. How they identified their "real self" increasingly moved away from institutional frameworks and toward more individualized and privatized criteria. Each person sought or was expected to work out his or her own self-definition in relative isolation from others. In a large empirical study, for example, Joseph Veroff and his colleagues, comparing the results of national surveys they conducted in 1957 and 1976, found a significant change in the way that people structured their self-definition and sense of well-being. They characterized this change as one from a "socially integrated" paradigm to a more "personal or individuated" paradigm and identified it in three aspects: a decline of social role standards as the basis for self-evaluation, an increased focus on self-expressiveness and self-direction, and a realignment of concern from social organizational integration to interpersonal intimacy.[25]

Along similar lines, social theorists have documented a decline of stable, role-based social relations and their mode of self-experience across many areas of social life. In his study of work in the "new capitalism," for instance, sociologist Richard Sennett stresses changes in the economy and the practices of the "flexible" corporation for creating new experiences of precariousness. Rigidity, stable routine, and hierarchy are out; dynamism, contingency, and network structures are in. "Workers are asked to behave nimbly," he writes, "to be open to change on short notice, to take risks continually, to become ever less dependent on regulations and formal procedures."[26] The new order of unrelenting change, whatever its economic consequences, fragments workers' experience of time, erodes those qualities of character that are long-term in nature, like loyalty and commitment, and challenges the ability of people to build or anchor a substantive and sustained sense of self. The new corporate form, he argues, demands a mobile sensibility—flexible, superficially cooperative, oriented to the short-term, and ready and willing to forget past experience and previous learning as changing circumstances require.

The radical flexibility and fluidity of new economic conditions are one source of change, but there are others, equally as radical. A wide range of social and technological transformations have destabilized many spheres of life and altered the relationship between time and space. Solid and enduring social structures, as Bauman observes, which gave time its historical and collective qualities of duration and rhythm and sequence, have grown ever more precarious or have disappeared. Time is now more like an "aggregate of moments," he argues, and there is a premium on movement and lightness over that which is fixed or stationary. Like Sennett, he is not suggesting that ordinary people live without any continuity or stability, but that social pluralization and destabilization are changing our perception and scale of values. Long-term thinking, planning, and acting, for instance, are increasingly difficult and devalued, replaced by the immediacy of short-term projects and episodes. Being bound to a place, once a revered condition, is now a sign of backwardness. Durability, once coveted, is replaced with transience. "Being stuck with things for a long time," he writes, "beyond their 'use up and abandon' date and beyond the moment when their 'new and improved' replacements and 'upgrades' are on offer, is . . . [a] symptom of deprivation."[27] What is valued today—by choice as much as by unchosen necessity—is flexibility, "a readiness to change tactics and style at short notice, to abandon commitments and loyalties without regret."[28]

Moreover, in addition to and in interaction with social and technological transformations, the pace of daily life is increasing. Technology allows us to do things more quickly and with less effort, which in itself should increase free time. But, according to German sociologist Hartmut Rosa, we now do more things in the same span of time, whether by doing them faster, compressing (shortening the time between) activities, or by multitasking. In other words, technology gains, such as from labor-saving devices, are more than lost due to increases in activity, activity that is connected with both technical and social change—toward greater output, improved quality, new patterns of relationship, expanded possibilities of action, and so on. The result, he argues, is a chronic shortage of time, an experience of being strapped with expanding obligations, options, and contingencies. And there is no resting place because of the rapid rate at which the reliability of experiences and expectations decays. We try and keep up or risk being left behind. Like the ceaselessness of economic production, in which standing still means losing ground, Rosa observes, any withdrawing or prolonged break from the ongoing change means missing out on potentially valuable options and connections. "Whoever does not continually readapt to the steadily shifting conditions of action," he writes, "loses the connections that enable future options." Without continuous striving in every area of life, we are constantly in danger of a status spoiling "anachronization."[29] "Everything's about going forward," according to another observer, "Falling back is the American nightmare."[30]

These studies and others like them document an increasingly uncertain and unpredictable social environment, the decline of institutional self-anchorage, and the devolution of choice and self-making to individuals. Any discussion of social change and its consequences is always in danger of overgeneralization and of missing alternative outcomes and important forms of resistance. There is no suggestion in this literature, however, that everyone is affected equally or similarly or that ascriptive identities (such as race and gender) and other fixities have all disappeared. Far from it. But what can be said of consequences, with considerable overlap in both theoretical and empirical assessments, is that the experience of self and consciousness has changed as less and less is institutionally provided to people and more and more is expected of them. These conditions promote a conception of ourselves as "the center of action, as the planning office with respect to [our] own biography, abilities, orientations, relationships, and so on."[31] They promote a strong form of "reflexive" self-awareness. This is a sensibility that is oriented to the expectations of others, acutely attentive to what is socially appropriate, and adaptive to the normative requirements of each context of interaction.[32]

It can also be said that in this individualized, fast, and mediated world, the personal risks have changed. They are not just greater in number, there are qualitative differences—the risks of chosen identity, of the consequences of decisions we ourselves have made, of events now regarded as personal failures, of missing out on the expanding number of realizable options.[33] Neither the world nor ourselves can be approached as given, determined, fixed. Rather, obstacles must be treated as "variables" that can be individually "moderated, subverted or nullified" by our creative efforts.[34] Certainly this new world has the virtues of dynamism, democracy, and the promise that we might become anybody, but it has many possible pitfalls. It is an environment likely to sap confidence, breed discontent, heighten unease and risk awareness, foster a frantic yearning for recognition, and generate a continuous sense of incompleteness and isolation. Just the sorts of experiences that participants described in their predicaments.

The sensibility and self-conception fostered by contemporary social conditions corresponds with features of the neurobiological imaginary. In particular, I want to briefly consider two areas of correspondence. One is how our mode of existence stimulates an outer rather than inner locus of subjectivity. This shift impacts how we think about our personal history and the point of reflecting on it. The second is how our mode of existence stimulates a less conflictual view of society than in earlier times. This shift impacts our awareness of and perspective on social norms.

### From Reflective to Reflexive

As much as any other force, technological change and acceleration have powerfully reshaped everyday experience. In *Mediated*, anthropologist Thomas de Zengotita, for instance, explores many features of daily life to show how our ubiquitous representational technologies, from television to computers to cell phones, have created a "society of surfaces." Our world, he argues, now comes to us thoroughly mediated, predigested, and packaged in images and addressed to us as an endless flow of flattering, experiential options—"your thing," done "your way." Negotiating the unceasing flow requires a "certain kind of very flexible self-awareness that depends on habitual reflexivity about emotions and relationships," he writes, and is a lot like method acting. Children come earlier and earlier to this intense self-consciousness, which is incompatible with the "stillness of depth" but is adaptive for maintaining a life of perpetual motion—"living in the moment," keeping options open, improvising social performances from a tool kit of adaptable postures.[35]

In our fluid, mediated environment, the premium is on a certain "lightness" of being—with flexible selves realized in a continuous foreground of movement and activity and acutely self-aware and sensitive to fitting in. What is valued is well described by sociologist David Riesman's classic ideal-type of the other-directed "social character."[36] This type, emerging in a consumer society, does not so much internalize a code of behavior when young—as with the earlier inner-directed type characteristic of a production-oriented economy—as a kind of sensory equipment. This elaborate equipment is needed to receive, and participate in circulating, the diverse and changing messages coming from a wide circle beyond the family, including peers, mass media, and consumer culture. For this type, the self-control mechanism is a form of diffuse anxiety that operates like "radar," sensing the signals—preferences, values, terms of approval—of each group milieu. These signals are the source of direction, used by the other-directed to adjust their internal states to match context-specific social expectations and the needs of the moment. Their locus of subjectivity, if you will, is on the "outside."

We can contrast this "light" mode of being with an ideal-type of persons whose locus of subjectivity is on the "inside," who seek to develop their inner world. Such a "depth model" once informed the picture of the human in many fields, including psychology and philosophy, and was also expressed in other areas of life, from art and architecture, to music and literature.[37] It is a model that presupposes a durable and tangible social world (visible authority, disciplinary norms, communal purpose). It presupposes a more enduring and rooted life experience, directed by an inner sense of purpose and a continuous narrative. It presupposes psychological complexity, with layers of meaning beneath the surface of our immediate awareness. It presupposes a critical attitude to the self-narrative, a painful awareness of the gulf between human aspiration and limitation, and a definition of selfhood in terms of tension and struggle both within (against temptation, instinctual drives) and without (against the constraints and distortions of society).[38] It is a model that is now hard to find.

If reflexivity is the type of self-forming activity characteristic of the other-directed, then we might use "reflection" to describe the sort of self-forming activity characteristic of a depth orientation. In the latter, self-formation is oriented to aligning one's actions with one's character and principles, with finding the means to ends that are at least in part given or viewed as obligatory. Think, for instance, of the confidential and introspective nature of the old practice of keeping a diary.[39] Contrast that with the public, reflexive self-work involved in maintaining a Facebook profile. If the diary engages and cultivates an interiority and draws on the past and fairly stable standards of judgment, Facebook requires and fosters the shaping of a moldable, outer self that is attuned to the present and normatively oriented to the self-chosen networks of relationships to which the medium itself directs and embeds self-experience.

In these remarks, I am not suggesting that we can simply deduce change to persons from social change or technological engagement. How we respond to change is far more complex and multifarious. But what I am suggesting is that our social and technological arrangements foster a mode of being, an orientation to ourselves as "light," as "laterally" rather than "vertically" oriented,[40] as, in effect, psychologically transparent. When everything is about going forward, when we try to live a faster life, when past experience becomes a mere drag, there is a powerful tendency to retract into the present. Under these conditions, there seems to be little time or warrant for the slow work of reflection. If conducting our everyday affairs encourages forms of control in reflexive self-possession and an instrumental stance toward ourselves, then concerns with our inner life, motivations, and evaluative stance would seem to be of marginal significance. Not to the actualization of our inner powers do we attend, but to reading and responding to social cues out in the world. If the ability to "move on" or "reinvent" oneself is expected and necessary, then there is little point to reflecting on one's history and on painful experience.[41] Reflection seems like a luxury, or worse, a form of "debilitating nostalgia."[42] When life is discontinuous, then perhaps it also plausible to imagine that parts of it can be hived off by an act of will.

### Self and Society in Harmony

Another important feature of contemporary life, already suggested by *Mediated* and the Facebook example, is that self-experience is increasingly located and embedded in networks of relationships. In a characteristic study with the revealing name *In Conflict No Longer: Self and Society in Contemporary America*, sociologist Irene Taviss Thomson shows how relationships have become "constitutive of the self."[43] This does not represent a return to role-based identities or mean that people are again embedded in local groups and existential communities. While Americans are organizing more, they are actually joining and participating less. Relationships are now something individuals "have," and many of these affiliations, as others like sociologist Dalton Conley note, are not centered in any local community but are "elsewhere," widely disbursed and maintained through electronic communications.[44] In our highly fluid and mediated social environment, the new relationality consists of individuals immersing themselves in unique and flexible configurations of groups and relationships. Though ready to disaffiliate as necessary, it is through these self-chosen groups and relationships that people fashion and anchor important aspects of their sense of self. This "relational self," Thomson argues, is other-directed, performative, and self-reflexive. A mode of subjectivity, to use my terms, that is flatter, more "exteriorized" and disbursed.

Compared to the days when identities were more anchored in institutional frameworks, relational selves, Thomson argues, have a far less constraining view of society. For this type, features of the social landscape appear mutable, an open field of possibilities, life chances, and potential (within reach) goals, rather than inflexible and restrictive. If one feels constrained, then the variables, as I noted above, whether situation or self, must simply be changed. With less visible tension and struggle, older distinctions based on self-society opposition, such as public and private, the individual and society, id and superego, authenticity and inauthenticity, lose their grounding. "Depth models" of subjectivity, as I called them, presupposed a divided self and a palpable social order that required that individuals be "hemmed in by disciplinary norms," to quote Ehrenberg again.[45] Alienation was once a common form of predicament; critics used to castigate social conformity. The relational or reflexive self, by contrast, in the act of authoring itself, scans its immediately apprehensible environment, highly attentive to the appropriate attitudes and behavior of its group milieus, and closely linking self-understanding with the right performance. The social, in this orientation, is a space of reflexive self-designing, not of resistance or repression.[46] "Conformity," in this plastic world, is not a critical evaluative concept, much less a shortcoming.

If the self-society tension is effaced, then everyday suffering must be conceived in other terms. When the dominant norms themselves become unlivable, throwing off suffering when we do not measure up or adapt readily enough or slow down instead of speeding up, the resulting predicaments (and sense of guilt) make no sense. The practice of reflexive self-making does not prepare a person for conflict in such terms. Instead, it is likely to foster a view of problems as self-inflicted or as completely opaque and mysterious, without rational explanation. In the first instance, there is shame and silence, and in the second, a search for a (nonself) dysfunctional mechanism. And the only logical response is to moderate, subvert, or nullify that dysfunction—to fix it—and to double down on mastering the norms so as to retake our proper place in the flow of things.

These brief reflections on contemporary social life are perhaps sufficient to suggest a homology between features of life today and what the interview participants said about the mechanical causation of their suffering, the type of hope they placed in medication, and the dismissive terms in which they spoke of insight and dialogical forms of predicament response. A neurobiological way of imagining ourselves is compelling because it is a sensibility promoted by the social organization and tenor of our times.

### Theodicy: Saved by Grace

By reconciling suffering, self-understanding, and social norms, the neurobiological imaginary represents a kind of secular theodicy. The problem of theodicy, of how to account for the fact that catastrophes occur and the good suffer, is ancient. The Book of Job and the flood narrative in the Epic of Gilgamesh are powerful examples. The modern concept dates to Leibniz and concerned the specifically theological problem of reconciling God's goodness and omnipotence with the existence of evil and suffering in the world. From this monotheistic context, the sociologist Max Weber extended the explanatory problem of theodicy to include a much wider range of cultural/religious efforts to account for misfortune and the "incongruity between destiny and merit." He grounded his expanded concept of theodicy in a philosophical anthropology that takes as basic the human need for a meaningful moral order to function in the world. This powerful dependency creates an "ineradicable demand" to make rational sense of evil, injustice, and suffering.[47] For Weber, theodicies serve a primary (though not exclusively) legitimating function. In the face of "irrational" experiences of suffering, they justify social arrangements and square people with their lot in the unequal "distribution of fortunes."[48]

The favored strata of society, those with social honor and power, Weber observed, tend to base their claim to status on a "special and intrinsic quality of their own," a feature of their "actual or alleged being."[49] He has in mind social arrangements like aristocracies with blood lineages, but, paradoxically enough, in our egalitarian and meritocratic system, status is based on similar claims. Status is challenged in predicaments for everyone because they involve norms of being. As I observed in chapter 5, with some clear exceptions, the operative norms of relationship, emotion, and self were quite similar for everyone, though there was some scaling of expectations depending on socioeconomic situation. But something like "actual or alleged being" based on one's own qualities was at stake.

Recall Hailey, who like many others, looks to the neurobiological imaginary and its mechanistic assumptions for compensation, as a way to close the gap between a viable way of being and perceived inadequacy. Against the unforgiving norms of being, she can affirm her good standing. This is the sense of relief, discussed in chapter 4, that people get from talk of diagnosis and neural mechanisms. They do not receive forgiveness for there is no one who can forgive them. Nor is magic at work, even if that is what many would hope for. Instead, what is received is a kind of grace. The word "grace" references God's assistance, but readily secularized, we see the same concept at work, of an assistance, in the words of Iris Murdoch, "to human endeavour which overcomes empirical limitations of personality."[50] The sufferer is symbolically restored to favor, when it seemed like the only possibility was in a lowered status of being.

This is a theodicy (or what some call a secular "anthropodicy") with none of the metaphysical assumptions about God or the world embedded in the "pure types" of religious theodicy detailed by Weber. Such larger sense-making frameworks have less and less purchase in our society. Yet, a secular theodicy responds to the same phenomenological need for order, explanation, and a reconciliation between normative ideals and lived experience.[51] As anthropologist Jean Comaroff writes, "We look to medicine to provide us with key symbols for constructing a framework of meaning—a mythology of our state of being."[52]

As the logic of neurobiological causation protects and affirms the being of the sufferer, it also protects and affirms the image of the liberal, self-defining self. It makes the norms appear realizable. And, at the same time, it defeats challenges and legitimates the normative order. In a secular context, irrational and anomalous experiences of suffering can call into question basic assumptions about personal freedom, social order, and the distribution of fortunes. In the face of predicaments, participants almost always questioned themselves, but in a few cases discussed in chapter 5, they questioned instead such standards and ideals as self-invention, optimization, and contractual relationships. These people did not describe their suffering in biological terms. For those who did, the medicalizing account has an answer, showing how incongruities that arise are not disconfirming of the liberal picture or social norms. Once interpreted as biological, their suffering has no real meaning; it causes pain, confusion, and inconvenience— but it *signifies* nothing. Their suffering does not reveal anything untoward about the regime of the self or contemporary social arrangements because it does not reveal *anything*. And the seeming proof, the evidence that suffering is a matter of individual neurochemistry, is medication.

## Conclusion

As a response to everyday suffering and the predicaments it generates, the neurobiological imaginary is compelling to think with. In the context of seeking to ameliorate or rise above a predicament, participants emphasized the autonomy, control, and efficiency of the viable self and sought to be less vulnerable and more self-sufficient, flexible, and optimizing. Moving toward viability—a strong version of the liberal ideal—involved getting free of encumberments on their volition. In its pure form, according to the anthropologist Deborah Gordon, the liberal self "strives to be its own author, consciously choosing its path, able to disengage itself and step back and judge rationally what it will be, where it will go." To this end, a disengaged perspective is necessary because, she continues, it is only by means of thinking of and enacting oneself in terms independent of any necessary reference points that one is "free to be one's own unique author."[53] Conceptualizing one's experience in *disengaged* terms—objectively, neutrally, instrumentally, noninterpretively—is the way to make the goal realizable.

This is the tendency we saw in participants like Hailey, the type of perspective that the neurobiological imaginary represents. Treating problems in a medicalizing framework makes them a technical problem, a malfunction in the individual. Linking limitations to the brain provides a way to make an object of that malfunction, an object that medication can then be interpreted as mechanistically removing. The strain between ideals and standard—a type of person, a way of being—and perceived insufficiency can be wiped away, as by grace.

It is not hard to see why such a perspective is becoming more and more common, and why the shift toward the neurobiological is so quiet. Much in our world promotes it. The imaginary can seem natural, in part, because we already adopt a disengaged stance toward ourselves in many areas of life and encounter it in social practices. It is not strange for us to think of our life as something we define by our choices or of the world as a space of phenomena that can be shaped according to our self-defined needs and purposes. Many of our practices stimulate a view of ourselves as "light" and psychologically transparent and direct us toward a way of conducting ourselves that is reflexive, present-oriented, fast-paced, and ready to change. In many contexts, there is a social premium on a self-mastery that consists in

treating ourselves as a kind of clay in our own hands, moldable and ma-
nipulable, confirmed by the right performance.

In the face of a predicament, that is the kind of self-mastery that the
neurobiological imaginary holds out. But just how much of ourselves must
we surrender in its pursuit? I conclude by considering the disengaged perspective and where it leaves us.

# A Crisis of the Spirit

Indeed, one might argue that psychopharmaceuticals have been more effective in persuading people of their essentially mechanistic and physical-chemical nature than all of modern science put together.

—Evelyn Fox Keller[1]

In interviews with ordinary Americans dealing with everyday suffering, I found a way of imagining the relation between self and body, suffering and society that I have called neurobiological. In interpreting their interpretations, I explored their thoughts, impressions, and assumptions within the larger healthscape that they drew upon, including medical language, intervention approaches, and popular sources of information. I set their evaluations and restorative ambitions against the deeper normative and social background that informs them, from conceptions of normalness/differentness to aspirations of viable selfhood. And I sought to show the social and moral appeal of the imaginary, what form of self-mastery it seems to offer for our liquid times.

In the course of the book, some troubling implications of imagining self and everyday suffering in this neurobiological way have also come up. These implications, as expressed by participants themselves, were both practical and existential. Their medication treatment might have to be lifelong, for instance, or they might require more than one drug, be dependent and unable to self-regulate, or in despair of ever (again) finding the right one. They might be different from other people. Some of their actions and emotions may be without reasons, not really theirs at all but with a separate ontology shaped by unknowable forces. Diagnosed with a mental disorder, they may be deprived of their vernacular language, feel the need to hide aspects of their experience, or resist the implication of loss of control by drawing a strong line of "otherness" around those with mental illness.

More could be said, but in these concluding remarks, I want to consider a number of broader possible consequences. The opening epigraph is from an essay by Evelyn Fox Keller on the diminution of first personhood. Of course, as I have tried to show, it is not the psychopharmaceuticals as such that have been persuasive but the psychopharmaceuticals as interpreted through the neurobiological imaginary. But the direction of persuasion, toward an "essentially mechanistic" view of ourselves, is just where the new imaginary takes us. It subserves a larger cultural aspiration of liberal selfhood, a promise of freedom, yet in a way that diminishes the person. To illustrate this propensity, I begin with cognate examples of the mechanistic and disengaged picture of the person in several scientific and scholarly fields. The alternative languages they speak in, I argue, fail as a more penetrating account of our situation and move toward the de facto authorization of a solitary and empty conception of the will. This is one impact of imagining ourselves through the prism of the neurobiological. There are other impacts, as well, on our relationship to social norms and ideals, our recourse to other evaluative frameworks, our self-understanding and relation to others, and our attitude to suffering. The imaginary does not open up interpretation, enable vision, or cultivate solidarity, but closes off and inhibits.

## The Disengaged Picture of Ourselves

We are not who we think we are. Or so we are told, informed that our understanding of ourselves as complex persons is fundamentally in error. The heralds of this pronouncement are remarkably diverse and come from regions of the intellectual world that engage in no direct conversation. Despite their seeming variety and even antipathic mentalities, they actually share a picture of the human with a distinct family resemblance. Each in its own way calls us to reject an internal, meaning-making, qualitative, first-person view of ourselves for some version of an external, disengaged, and mechanistic perspective in which we think of ourselves or crucial aspects of ourselves independently of our intentions, our inner experience, our evaluative outlook, and our place in the social and cultural world.

Many works of popular genetics and neuroscience present the new picture in the context of spirited polemics against the mind and soul. They resolve the old mind-body problem by reducing the mind to the brain and treating the distinctively human qualities, such as reason, thought, and a moral sense, as epiphenomenal manifestations of more fundamental natural mechanisms and processes—genes, hormones, neurons. "Humans possess no special capacities," according to one of these books, "no extra ingredients, that could conceivably do the work of the mind, the soul, or free will as traditionally conceived."[2] Consciousness, intentionality, and all the familiar features of our subjectivity are actually brain functions, constructed by the brain and projected onto the world. "'You,'" according to geneticist and Nobel laureate Francis Crick, "your joys and your sorrows, your memories and your ambitions, your sense of personal identity and free will, are in fact no more than the behavior of a vast assembly of nerve cells and their associated molecules."[3] It is the brain, according to the neuroethicist Patricia Churchland, in a book with the subtitle *The Self as Brain*, that "holds the key to what makes me the way I am."[4]

Similar claims, including the rejection of free will and what philosophers call the first-person perspective, also appear in works of cognitive and behavioral science. Summarizing some of this research, one philosopher concludes that the "upshot of all these discoveries is deeply significant, not just for philosophy, but for us as human beings: There is no first-person point of view. Our access to our own thoughts is just as indirect and fallible as our access to the thoughts of other people. We have no privileged access to our own minds." And, therefore, "introspection and consciousness are not reliable bases for self-knowledge."[5] A similar conclusion comes from parts of social and experimental psychology, which have come to be heavily influenced by the "automaticity juggernaut." In this view, everyday thought, feeling, and action is in large measure, if not almost wholly, controlled by reflex-like brain processes that operate outside conscious awareness or voluntary control.[6] Little in the mind is even available to introspection; what moves us are forces that we, by ourselves, have no power to reflect upon, interpret, or understand. We are "strangers to ourselves."[7] If you really want to understand yourself, convene a focus group. Much contemporary philosophy shares a nearly identical perspective.

The same disengaged picture can be found elsewhere and in contexts far removed from talk of the brain. In the poststructuralist, postmodern, and cultural studies wings of the humanities, you will find a remarkably similar injunction to abandon the languages that we have traditionally used to make sense of ourselves. In this literature, the modern rational and unified subject is a myth, produced by systems of domination, exclusion, and control. In its place is an image of the self in pieces, a decentered and discontinuous assemblage of experiences. "Identities," according to the late social theorist Stuart Hall, "are points of temporary attachment to the subject positions which discursive practices construct for us."[8] The "inner" lacks any stable substance or orientation; the self is externally determined, merely an effect of roles and discourses that are instrumental, performative, and shaped by systems of power.[9] As they harden over time, they produce the misconception that we have a self on the "inside."

All these examples share a deflationary, debunking principle, demanding that we muster the courage and honesty to recognize the illusory character of the persons, with souls and minds and depths, we thought we were. At the same time, they are also carried on, despite the seeming determinism (whether biological or social), in an optimistic spirit. The polemic against the inner life, the overcoming of subjectivity, is somehow required to clear the way for progress. For the scientists, biology need not be destiny. In their literature, research advances in genetics and molecular biology, neuroscience, evolutionary psychology, and other fields, along with their respective technologies, will give us, to paraphrase Churchill, the key to what makes us the way we are and open up new possibilities for intervention. For the poststructuralists and other theorists, freedom from the modern language of rational, centered selfhood promises to be liberating. The protean, fragmented conditions of contemporary life open up new possibilities to situate new subjectivities and remake the social order in ways that are only beginning to be defined.

Further, despite the putative determinism, advocates of the external perspective smuggle human agency and the liberal self back in. The activists in the sciences, like Richard Dawkins and Steven Pinker, carry on in this way, as did such leading postmodern figures as Michel Foucault and Richard Rorty. Each keeps in their picture of the human certain critical capacities, such as the ability to resist domination, that are incoherent without an acting person. And, in purely logical terms, any attempt to rescue us from our illusions, much less specify progress and liberation, must include, however unacknowledged, concepts of truth and freedom and some notion of human agency to move toward them. In an effort to have it both ways, some advocates have suggested that while self and personhood and free will are illusions, they are useful illusions.[10] We have no soul, yet life is more meaningful and manageable when we operate as if there is in fact something more.

Practically speaking, some such equivocation seems inescapable. Taken to its logical conclusion, the external perspective would leave us, to quote the philosopher Thomas Nagel, with "no one to be."[11] Further, because they do not reveal us, our actions would become mere events and other people mere things. But the scientists and the professors do not want to argue us out of existence, were that even possible. Instead, and paradoxically, they place us in something like the position that philosopher Iris Murdoch once discerned in moral philosophy. In the approaches dominant in philosophy, she observed, what we are "objectively" is not under our control, while what we are "subjectively" shrinks to a "solitary, substanceless will."[12]

## Languages of Selfhood, Solidarity, and Suffering

For all their seeming self-assurance, neither the sciences nor the postmodern humanities can explain the mind or the self in anything like the way they promise. We are flesh-and-blood creatures to be sure, but we are not our brains, and our consciousness, our minds, our selves cannot be comprehended in the mechanistic language of neurobiology.[13] In fact, nothing of everyday suffering is even visible in that language. Nor does poststructuralist theory with such notions as radical self-making and vitalism actually provide a new way of being or an alternative language. Some have suggested, for instance, that imagining suffering in terms of the neurobiological might provide a language to break from particular, communal pasts, such as the old conventions of mental illness narratives for women.[14] Others have suggested that the turn to neurobiology represents a shift toward an intensified identification with the body and embodiment or a breaking down of the old Cartesian dualism between privileged mind and devalued body. The sociologist Nikolas Rose, observing the very trend I am describing—that is, to map what would have previously been seen as "psychological" discontents onto the body—makes such an argument, that we are coming to see ourselves as "somatic individuals."[15]

Participants drawing on the neurobiological imaginary to explain everyday suffering did not speak of any increased identification with the body and, for the most part, retained the language of liberal selfhood in evaluating their experience and articulating the ideal toward which they were aiming. Appropriation was on their own terms, at a distance from notions of illness and mental disorder, and the fumbling talk of neurochemistry was not a means to frame the concept of selfhood. It was deployed to split off limitations from the self, still conceived as the conscious seat of individuality and control; in their stories, the neurobiological was a container for the "nonself."

These languages do not provide a deeper, more penetrating characterization of everyday suffering—of hopes and fears, desires and ambitions, needs and losses, limitations and inadequacies—or of social circumstances.

Just the opposite. What, after all, would it mean to say, as many participants did, that whole areas of thought, feeling, or behavior are chemically determined, alien to their social and dialogical history, life experience, personal concerns, and commitments? Would it even be possible to detach so much of one's evaluative outlook and relation to self? With a medicalizing perspective, troublesome experience is flattened out, reduced and recast in an impersonal language of symptoms, diagnoses, and brain states; it is rendered illegible. Instead of greater articulateness, it brings the process of interpretation and self-clarification to an end. The antihermeneutic, postmodern celebration of the "surface" does the same.

The medicalizing perspective does not gain its plausibility from any power to explain the dilemmas of selfhood or disclose the phenomenological world of everyday life. In fact, it sweeps all that aside and makes inquiry impossible. As the scientists and philosophers have reminded us, the brain and brain processes are simply not accessible to self-exploration or interrogation. Toward the "something biological" of which participants spoke, there can be no first-person access whatsoever. The cause of suffering is a force that we, by ourselves, have no power to reflect upon, interpret, or understand. It is utterly beyond our reach. We can have no sense of what this force might even look like. For what we would discover if we did have access is not a desire, or a belief, or an intention or anything of the sort, but a process, a brain system activated by stimuli, neural firings in themselves equally without content, or substance, or meaning. Setting aside social and dialogical aspects of selfhood consigns crucial experiences to silence.

Further, and for the same reason, the neurobiological imaginary embeds participants more firmly in the regime of viable selfhood and tightens its hold. Since the norms are built directly into the medicalized language, any recourse to that language cannot but reify the social norms as the natural and inevitable yardsticks of health and lead to a weighing of experience in a fixed relation to them. No other normative standard can be brought to bear since the very presence of a social standard is a priori denied. Medical language is, by definition, a value-neutral matter of the body, indifferent to ideals or social standards or different conceptions of the good. The existential question of what it is good to be is already indicated, calibrated (with each version of the DSM, for instance, or in clinical encounters) to the shifting norms of being. The only questions are technical, about the means to being "workable" and ideally enjoying an "undisturbed self-existence" in the world.[16] There is no distance here from the norms or grounds on which to criticize the regime itself or build solidarity or break free of the perpetual sense of deficiency it fosters. There are alternative norms and alternative perspectives on our predicaments, personal and social, but they are not visible within the neurobiological imaginary.

### *Subjects Not Objects*

There is a reductionism here, but it is not a biological reduction. In endorsing the dimensions of the neurobiological imaginary, the people we interviewed were not endorsing some determined picture of themselves, without free will or effective agency in the world. Again, just the opposite. As we saw in chapter 4, they expended considerable energy to defeat implications of being determined or different. Much like the purveyors of the disengaged and external perspective in the sciences and philosophy, they did not draw on biology to replace the liberal autonomous self and ideal of self-mastery but to further secure them.

The reduction here is of the human person. In psychosocial approaches, the self emerges in social interaction and practices, is constituted through ordinary language and narrative (teleological and intentional), and is articulated through personal engagement and reflection on experience and one's position within a shared symbolic universe. Similarly, in the hermeneutic approach in the social sciences and philosophy that I have been working from, our acts (inner and outer) and our embodiment belong to our "continuous fabric of being."[17] Movement toward a richer and more accurate understanding of ourselves involves our temporality (our memory and history) and the social practices and dialogical relations—with ourselves (internal conversation), with others, within a community and a tradition—that have helped to make us who we are. Self-elaboration, in this view, is an ongoing ethical activity, an effort to see ourselves clearly, to gain a fuller picture of who we are, what moves us, and—critically—by what standards we live. And this elaboration can serve as a basis for the cultivating of sociality and solidarity with others.

Movement toward a richer and more accurate understanding can involve struggle, the help of others, and a checking of our interpretations against the facts of our own history. It requires reflection, in the sense discussed in chapter 6, on our conduct, thoughts, and feelings. Our emotions, the social emotions that involve our relations with other people, like shame, remorse, pride, admiration, envy, jealousy, self-contempt, disappointment, and guilt, are especially crucial to self-knowledge because they are relational to, and in a feedback loop with, what we care about.[18] They concern our status and involve our subjective compliance and self-worth.

They are emergent from situations and provide, in the words of Margaret Archer, "commentaries upon our concerns."[19] The commentary is an evaluative awareness of a situation and its relevance for our desires and aversions, attachments and aspirations. To experience an emotion is to perceive a situation as bearing a particular import for us and setting the stage for potential action.

Of course, how we evaluate the situation may be wrong or unwarranted, and so we "can be mistaken," to quote Iris Murdoch again, "about what we think and feel."[20] Our feelings may be at odds with our understanding of the circumstances, or we may misapply norms to ourselves—such as when we feel responsible for things completely outside of our control. We may sense an emotion is prohibited and seek to redirect its import to something less threatening.[21] The fallibility of our emotions is an important reason why reflection is necessary. On further consideration, we may reject previous interpretations as incomplete or wrong in our effort to elaborate a more accurate characterization of what a situation meant to us. The new formulation, in turn, can and often does, lead to changes in how we feel.

For participants, this second-order (introspective or dialogic) reflection is precisely what the neurobiological interpretation cuts short or eliminates. Recasting emotions as objectless, as the result of rogue chemicals, negates further reflection and understanding and ends any reassessment of self or circumstances that might have followed. Like those who ended psychotherapy after beginning medication said, there was nothing further to talk about. Similarly, once people adopt disorder categories, like depression or SAD, to characterize their emotional responses, there is a strong tendency to then use the new language to displace any richer emotional vocabulary. In many of their stories, for instance, participants reduced every negative feeling to such blanket terms as being "depressed" or "anxious." These terms are not an alternative way of expressing the emotions the participant experienced; rather, the new terms reorder and homogenize what they experienced. They bring a different reality into being, altering the fields of meaning. Once altered, the specific import of participant emotions for self-understanding is undone. Their original experience, which would be critical for understanding what was going on, can no longer speak to them.

The neurobiological imaginary also walls off or drives out other evaluative frameworks. Transforming existential questions into technical questions not only supplants perspectives and interventions that presuppose the inner life, as we have seen, but it censors everyday sense-making. This may be a reason why changes of mind among participants only went in one direction. Perhaps once experience has been recast in a medical episteme, there is no going back.[22] Irrational and puzzling thoughts, actions, and feelings can no longer be perceived in the everyday explanatory language of beliefs, desires, hopes, and fears. Now the vernacular seems insufficiently materialist, subjective rather than objective, insubstantial not real, victim blaming rather than value neutral. Against the prestigious ontology of science, in short, it seems invalid, incompetent, even mean-spirited. Just such a loss of confidence in their own internal resources, already promoted by the healthscape, is further reinforced.

A reliance on expert language (even as filled in with personal meaning) changes relations with others. It demotes those who cannot operate in that language, reducing the role they and any supportive community might play. At best, as we saw with the role of confidants, they can recommend or confirm the need for expert help and affirm the patient in complying with it. But through the medicalizing lens, this route verifies for patients that their suffering is in fact a private burden, a somatic dysfunction in them for which others' effort and responsibility to care will be sharply circumscribed. Others cannot help them explore alternative methods of coping or work their predicament out. Beyond the apprehension about being regarded as different, this marginalizing of support contributed to the feelings of isolation that many participants spoke of. The nature of the suffering, as such, could not be voiced. Even the professional, to the surprise and frustration of some participants, focused on symptoms to the exclusion of them as suffering persons. A larger discussion of their situation was, as a matter of course, not permitted. Friends and family, knowing of their treatment, also appeared to offer them less leeway, quick to associate any changes in behavior or mood to their medication.

Finally, thinking in the neurobiological imaginary can strip suffering of any meaning and therefore of any value. Especially those with a medicalizing perspective could find no significance or purpose in their struggles or grounds for valuing aspects of themselves that are at odds with the norms of being. They did mention increased empathy for others, but this was in the narrow sense of a greater recognition that "everyone has issues." They did not speak of deeper relationships forged in the furnace of the predicament. The one lesson they could draw from their experience was that they should have sought to get the broken mechanism fixed earlier. There are long religious traditions and a large body of literature, from Sophocles to Dostoyevsky to Edith Stein, that see in suffering the possibility to enrich the soul, gain clarity, and apprehend important truths about ourselves, the human condition, and our bond to others. Such talk is from different and richer imaginaries, its very foreignness illustrative of the shift that I have sought to describe.

The shift toward the biological imaginary has its reasons, its appeal, and its promises. It comports with many features of our life today, and in many contexts, it is esteemed as a sophisticated attitude and a "best practice." And since the norms of being seem like good things in themselves—living up to one's potential, being all you can be—it is hard to see how their pursuit might lead to suffering. At the same time, thinking in the neurobiological imaginary blinds us to the social dimensions of our struggles, silencing the nature of what troubles us, and implicitly underming important truths about our moral freedom and the place of self-knowledge in a well-lived life. It obscures the whole person who suffers with a shallow, mechanical view of human being that impoverishes us all.

ACKNOWLEDGMENTS

The debts I have run up, large and small, in conducting the research and writing of this book have been many and far exceed what I can adequately acknowledge. My thanks, first and foremost, are to Susannah Myers, who— while studying law—served as research assistant extraordinaire. She was simply vital to every aspect of the project, from background research to copyediting to hashing out interpretations. To her, my deepest gratitude and appreciation. Several colleagues helped with conducting interviews and background research including Teresa Doksum, Lisa Magged, Sarah J. Shoemaker, Christina Simko, and Regina Smardon. I am grateful to each for their careful and thoughtful work. Among other assistance, Joshua Caler and Ben Snyder read transcripts of some of the initial interviews and participated with me in something like group grounded-theory sessions. Our goal was to ask, at a most general level, what the interview participant was saying and to form initial impressions. My thanks to them both for these fruitful discussions and for all their other input and suggestions.

A number of colleagues convened in Charlottesville in December 2017 to undertake a critical reading of the manuscript in its initial draft. To Charles Bosk, Carl Bowman, Tal Brewer, Matt Crawford, Carl Elliott, and Allan Horwitz, on whom I imposed this workshop, I am deeply indebted. The book, whatever its shortcomings, is far better because of their critical feedback. I also owe special thanks to David Karp, who shared with me the interview guide he used in his study *Speaking of Sadness*, and offered very helpful advice on how to let conversations unfold when speaking with people on sensitive, personal matters. He is a master of the art.

Doug Mitchell, sociology editor at the University of Chicago Press, was, as always, an inspiration. In his last fall at the Press before retirement, he took the manuscript through the review process, and did so with the intellectual generosity and collegial good cheer for which he is justly famous. Another consummate editor, Elizabeth Branch Dyson, picked up the project for the next phase and was a delight to work with. Her enthusiasm for the book is just what every author craves. The external reviewers for the Press were an immense help. I am especially grateful for Owen Whooley's detailed and beyond-the-call-of-duty review. His penetrating comments provided a virtual blueprint for the final editing of the manuscript and the book was greatly improved by his careful reading.

A number of colleagues, past and present, have been invaluable interlocutors. I especially mention Tal Brewer, Matt Crawford, Bill Hasselberger, Justin Mutter, James Nolan, Paul Scherz, and Jay Tolson. I have long benefitted from the unique intellectual environment of the Institute for Advanced Studies in Culture here at the University of Virginia. My thanks, as always, to James Davison Hunter and to Ryan Olson. At Macalester, where I wrote for several summers, Brooke Lea and Jaine Strauss were kind and generous hosts. I was fortunate to have funding in the early stages from the Pew Charitable Trusts.

I am indebted to the many people who agreed to be interviewed and who shared such intimate details of their lives. I hope that what I have written here faithfully reflects their stories and repays their trust. Special thank as well to the students in my Prozac Culture seminar, where many of my interpretations began to take shape.

Finally, I dedicate this book to my wife, Monica. This book has occupied me for years, and while writing a book on everyday suffering, I fear that I inflicted more than a little on her. No words of gratitude are sufficient to express what I owe her. But I think she knows.

## Interview Sample

This is a book about personal experiences of suffering, conveyed to me through interviews. In order to recruit participants, I posted advertisements on the internet site Craigslist, placed notices in alternative weekly newspapers, and posted flyers on college campuses, other public kiosks, and public buildings such as libraries. The ads asked potential participants if they struggled with being sad, with being anxious in social situations, or with concentration and attention problems and whether they might be willing to talk about their experience. Because the interviews would last two hours and the participant would have to travel to our offices, I offered them a fifty-dollar gift card to participate.

Those who responded to the ad, ages eighteen and over, called an 800-number and were prescreened for eligibility. Exclusion criteria reflected the emphasis on those who were not suffering from serious mental illness. I did not interview anyone who had ever been hospitalized for a psychiatric condition or was taking an antipsychotic or other medication normally intended for those with a serious psychiatric condition. I did not interview people who were suffering from such illnesses as bipolar disorder, schizophrenia, or PTSD. Overall, because of these criteria, I did not interview more than half of all those who responded to my ad.

In recruiting participants, I selected for a purposive sample of the broader American public. A purposive sample is not random or representative. Rather, people are sought who constitute a diverse sample of the society from which they are drawn (in this case, residents of the United States) and whose experience can illuminate the key concerns and topics of interest. As noted in the introduction, I recruited and interviewed people in Chicago, Baltimore, Boston, and central Virginia (Charlottesville and the more rural Harrisonburg) in order to get some geographic representation. To ensure diversity, I sought to get some balance of men (one-third of the total), members of minority groups (Asian, Hispanic, or African American—one-third of the total), and, though not in specific proportions, individuals of both lower and higher socioeconomic status.

A key theme of the study concerned the experience of taking a psychoactive medication. To better understand the role of this experience, I needed a comparison group of people who shared similar sorts of struggles, with or without a diagnosis, but were not taking a medication. In the final sample, forty-three were on a medication, and thirty-seven were not. Although I originally planned to interview one hundred people, I stopped at eighty because I felt I had reached "saturation," meaning that further interviews would only confirm the principal patterns (discussed in chapter 1) I had already found. Many other interviews, not formally represented here, done by students for their research projects and in my classes, also suggested these patterns.

In its final form, the sample had two unanticipated characteristics that proved to be very important. First, by their wording and placement, my ads seem to have drawn out people who had at least some experience of psychotherapy or other form of counseling. Indeed, more than half of those interviewed had a history of counseling, which in some cases was ongoing at the time of the interview. This substantial percentage allowed for a meaningful comparison of perspectives on talk therapies and medication.

Second, most of my study participants who had seen a doctor consulted a psychiatrist. By contrast, studies show that some 70 percent of psychotropic prescriptions are written by nonpsychiatrist physicians. It is frequently claimed in the clinical literature that these regular physicians have little training in matters of mental health, are too quick to diagnose and prescribe, and may not inform patients of other treatment possibilities. My concern here is not with casual prescribing, a practice for which there is already abundant evidence. But if there is truth in the claim that regular physicians, as compared with psychiatrists, are not making patients aware of the range of treatment possibilities, then the experience of those I interviewed would reflect better-informed and more careful clinical practice. The fact is, whether at the suggestion of a psychiatrist or not, a lot of those interviewed did take advantage of a nonmedication treatment option. Their experience and views are thus a more reliable guide to the substantive trend I see taking place. Their experience also means, though, that I am not dealing with the persons most casually diagnosed and treated.

## The Interviews

The interviews, conducted by myself and research assistants, were structured by an "interview guide," one for those on a medication and a slightly different one for those who were not. The guides outlined the questions to be asked and possible prompts and follow-up questions. Interview guides are preferred in this type of research. They structure interviews but also allow participants room to tell their story and propose new lines of discussion.

We began each interview by asking the participants to describe the circumstances that brought on the "suffering"—actually, what "issue" in my ad they were responding to—or, in cases where no specific circumstances could be identified, the point at which they first felt something was wrong. We asked in what terms they described the "issue" (and then used that language) and how it has affected them. We asked if they had sought medical, psychotherapeutic, or other help, and, if so, about their interactions with professionals, if a diagnosis had been made or a prescription written, and what they learned. We asked their view of the cause of their suffering, sources of information they pursued, and whether their understanding had changed. We also asked whom they had told about the problem and whom they would not tell, and why. Next, we asked about what steps they have taken or are taking to try to ameliorate their predicament or rise above it. For the majority (two-thirds of the total), these steps included some professional help. And, finally, we asked them about the role of personal struggles in life and whether their experience had led them to revise the hopes and expectations they have for themselves. The interviews were digitally recorded and transcribed for analysis.

## Interpretive Method

The interview questions were organized in a sequence that aimed to draw out a story. Conventional qualitative methods in sociology involve data analysis in terms of concepts, either brought to the research from specific theoretical models or from in vivo categories that arise inductively from the data itself. This is certainly valid, and I drew upon grounded theory methods that involve the inductive approach. However, the emphasis on concepts typically involves breaking up interview data into separate speech sequences, comparing them across cases, and then illustrating the resulting patterns with brief quotes. The problem is that the quotes lose some of the very context that is the point of face-to-face interviews and their advantage over other methods.

In my analysis, I treated the whole interview as the unit of analysis, looking for patterns within each interview and only then comparing these patterns across cases. For example, we asked people who had been to a doctor to talk about the visit in which they received a diagnosis. Standard approaches would be to compare the answers that different participants provided to this question alone. But what participants said about doctor appointments is not some recitation of neutral facts or independent of what they said about other aspects of their predicament and must be interpreted as such. This is what narrative analysis seeks to do—preserve context by seeing all of the elements of a story in their relation to each other and to meanings in the wider society. Consequently, I first asked how what was said about the doctor visit related to the story as a whole and only then compared it to other stories and identified patterns. The writing was therefore organized in terms of groups of participants rather than illustrative quotes.

# NOTES

INTRODUCTION

1. Except where brand names are important to the story, I will avoid their use and simply refer to drugs by their class, such as antidepressant or stimulant.
2. All names for the interview participants are pseudonyms. In some cases, minor details in participant stories have been altered to protect anonymity.
3. Psychoactive medications, sometimes also referred to as psychotropic or psychiatric medications, act directly on the central nervous system and are prescribed to change mood/affect, cognition, and behavior. The general categories are antipsychotics, antidepressants, psychostimulants, anxiolytics or tranquilizers, mood stabilizers, and (so-called) cognition enhancers.
4. Though not formally represented here, the interpretations I offer have also been informed by more than a hundred other interviews conducted by students in research projects and my classes over the past ten years.
5. Kessler et al., "Prevalence and Treatment of Mental Disorders."
6. While the extent of the divergence between practice and strict medical criteria is a matter of debate among professionals and in the popular press, the fact of a divergence is widely acknowledged. This is particularly clear in the debate in the psychiatric literature over whether treatment should be concentrated on serious disorders or extended—as it is in practice—to "mild disorders" and "subthreshold syndromes." See, for example, Kessler et al., "Mild Disorders Should Not Be Eliminated from the DSM-V."
7. Horwitz, "Transforming Normality into Pathology."
8. Healy, The Antidepressant Era.
9. Lane, Shyness.
10. Visser et al., "Trends in the Parent-Report."
11. Carlat, Unhinged.
12. Kessler et al., "Prevalence and Treatment of Mental Disorders."
13. Malpass et al., "'Medication Career' or 'Moral Career'?"
14. DSM-5, 20.
15. To home in on this precise group, I used some rough criteria to exclude those with serious mental disorders. See the appendix for further discussion of my methodology.

16. For a historical overview of the literature, see Davis, "Medicalization, Social Control, and the Relief of Suffering."

17. Lane, *Shyness*; Conrad, *The Medicalization of Society*; and Frances, *Saving Normal*.

18. There are certainly exceptions to this generalization. One of the best is Horwitz and Wakefield, *The Loss of Sadness*. David Karp's interview-based books on depression and medication are also exceptions, though his participants are people coping with fairly severe major depression. In his first book, *Speaking of Sadness*, three-fifths of the participants had been hospitalized at least once. In his second, *Is It Me or My Meds?* (17), he observes that 40 percent of his sample had spent time in hospitals and that his sample is "skewed toward those who are more severely ill than would be the case for all Americans treated with psychiatric medications."

19. See Moynihan and Cassels, *Selling Sickness*, and Payer, *Disease-Mongers*.

20. Here is a particularly clear example, from the *New York Times*: "Valium was, significantly, one of the first psychoactive drugs to be used on a large scale on people who were basically fine. . . . These were the drugs that gave us a new way to slay our inner demons, medicating our way to a happier life." Henig, "Valium's Contribution to Our New Normal."

21. The notion of "thick" comes from Geertz, *The Interpretation of Cultures*. The sociologist Isaac Reed (*Interpretation and Social Knowledge*, 23) speaks of "maximal interpretation" and places it along a continuum with "minimal." While the minimal uses theory lightly, "The maximal end of the spectrum involves statements that mix, in a consistent and deep way, theoretical and evidential signification, in an effort to produce a powerful comprehension of the matter at hand." That is the degree of interpretation I'm after here.

22. In a world where so much that is human has been medicalized, the French sociologist Alain Ehrenberg (*The Weariness of the Self*, 6) asks, "Can we still make the distinction between the unhappiness and frustrations of everyday life and pathological suffering?" He is right to be skeptical, and at this late date, the very idea of boundaries seems hopelessly lost. But as "mental disorder" has come to have a particular somatic meaning in the DSM (see chapter 2), the distinction is even more important.

23. See, for example, Berger and Luckmann, *The Social Construction of Reality*.

24. Archer, *Being Human*.

25. Wilkinson, *Suffering*, 16–17.

26. The physician Eric Cassell ("Recognizing Suffering," 24) defines suffering as "distress brought about by the actual or perceived impending threat to the integrity or continued existence of the whole person," noting how a key aspect of the whole person is his or her purposes or aims.

27. See Davis, "Adolescents and the Pathologies of the Achieving Self."

28. Kierkegaard, *The Sickness unto Death*, 19, original emphasis.

29. The concept of "norms of being" comes from Erving Goffman (*Stigma*, 128). He uses the concept mainly with reference to socially valued bodily attributes, everything from race and age, to height, weight, and "physical comeliness." My use of the concept here is broader, to valued norms of self, emotion, and relationship that show one to be a particular type of person—what I will call, based on participant comments, a "viable self." The fuller discussion is in chapter 5.

30. See Taylor, *Sources of the Self*; Sayer, *Why Things Matter to People*.

31. According to the physician George Engel ("The Need for a New Medical Model," 130): "The more socially disruptive or individually upsetting the phenomenon, the more pressing the need of humans to devise explanatory systems."

32. Connolly, "The Human Predicament," 1121.
33. On the nature of accounts, see Davis, *Accounts of Innocence*.
34. Of the eighty total interviews, thirty expressed a predicament in terms of performance, eighteen in terms of loss, and twenty-three in terms of achievement. In four cases, participants did not regard themselves as having a predicament. They are not included here or in any of the analysis that follows. And there were five cases that could not be reliably classified into one of the three general predicament categories.
35. A common strategy of cultural analysis is to explore "marginal" experience, like illness or deviance, because it reveals deep commitments, tacit norms and rules, and the symbolic boundaries that mark off the context in which meaningful thought and action can take place (see Wuthnow et al., *Cultural Analysis*, 260). Suffering is an especially revealing marginal experience, according to pioneering sociologist Max Weber, because it is at the very heart of our meaning-making in the human world.
36. In his aptly titled *Acts of Meaning*, the cultural psychologist Jerome Bruner argues that in an interpretive social science, we are "trying to understand how human beings interpret their worlds and how *we* interpret *their* acts of interpretation" (xiii, original emphasis).
37. On seeing interviews as narratives, see Luker, *Salsa Dancing into the Social Sciences*, 167.
38. *Culture and Truth*, 16, quoted in Sayer, *Why Things Matter to People*, 3.
39. This is a moral sociology, a sociology that takes "moral phenomena" seriously. See Boltanski, *The Foetal Condition*, 234. On the inescapable importance of the moral dimension of human life, see Smith, *Moral, Believing Animals*, 8–20; and Taylor, *Sources of the Self*, chapter 1.
40. For a careful philosophical explication of "tacit" or "background" knowledge, see Collins, *Tacit and Explicit Knowledge*.
41. Reed (*Interpretation and Social Knowledge*, 104–5) writes, "Reconstructing the meanings of social life that are not consciously intended or obviously denoted by interviewees' statements or archival documents, but that underlie, form, and structure those very statements, is the primary intellectual goal of the interpretive epistemic mode."
42. Kadison and DiGeronimo, *College of the Overwhelmed*, 37 (emphasis removed).
43. For other examples, see Hara Estroff Marano, "Crisis U," *Psychology Today*, September 2015, https://www.psychologytoday.com/us/articles/201509/crisis-u; and Greg Lukianoff and Jonathan Haidt, "The Coddling of the American Mind," *Atlantic*, September 2015, https://www.theatlantic.com/magazine/archive/2015/09/the-coddling-of-the-american-mind/399356/.
44. Kendler, "Explanatory Models for Psychiatric Illness," 695.
45. See, for example, Ahn, Proctor, and Flanagan, "Mental Health Clinicians' Beliefs"; and Haslam, "Folk Psychiatry."
46. See, for example, Taylor, *Modern Social Imaginaries*.
47. I take the concept of a healthscape from Clarke, "From the Rise of Medicine to Biomedicalization."
48. See, for example, Davis, *Accounts of Innocence*.

CHAPTER ONE

1. I take these terms from the "folk psychiatry" framework of social psychologist Nick Haslam, but I am defining them in terms of what I heard from interview participants. Haslam, "Folk Psychiatry."

2. By lay viewpoints here, I am referring to individuals not groups of laypeople who are collectively organized, such as into patient advocacy groups. These groups have a different dynamic. As Brian Wynne ("Knowledges in Context," 115) writes, "Organization allows more comparison of experiences and expert accounts, more accumulation of alternative perspectives and questions, and more confidence to negotiate with or challenge imposed frameworks."

3. In addition to the common "deficit model," which treats the inevitable difference between public and professional understandings as a sign of lay ignorance, there is another common approach that treats lay views as simply wrong and as not constituting real knowledge. In this approach, the goal of investigation is to identify the social determinants that "could," to quote sociologist Jeff Coulter ("Beliefs and Practical Understanding," 165), "be claimed to engender the false beliefs." A method, he continues, that deprives lay beliefs of "consideration in terms of the reasoning which may inform them."

4. There was a time when Americans would have included clergy members among those they might seek for counseling about everyday suffering. See Veroff, Kulka, and Douvan, *Mental Health in America*. Among participants in my study, only one spoke of counseling with a church pastor. The others referenced various psychologists, social workers, and doctors.

5. Several research studies have suggested that people might take on a chemical explanation simply to more readily justify a treatment preference for drugs. See Prins et al., "Health Beliefs," 1047; and Goldstein and Rosselli, "Etiological Paradigms," 560. The evidence I found suggests that taking on a chemical explanation is actually quite challenging and in a feedback relationship to medication-taking.

6. Frank and Frank, *Persuasion and Healing*.

7. It is very common for psychotherapy clients to also use medication. According to a study of national trends, more than two-thirds of those who received psychotherapy in 2007 were also taking a psychotropic medication. Olfson and Marcus, "National Trends in Outpatient Psychotherapy."

8. The fact that most participants had seen two different professionals reflects the decisive shift by psychiatrists away from practicing psychotherapy and concentrating solely on medication treatment. Mojtabai and Olfson, "National Trends in Psychotherapy."

9. In a study published in 2003, Iselin and Addis ("Effects of Etiology") found that biological attributions of problem origin reduced the perceived effectiveness of psychotherapy. See also Schreiber and Hartrick, "Keeping It Together."

10. An interesting contrast is Donna, age twenty-five, who has had a series of struggles with self-image and eating issues dating back to the time in her teen years when her father was imprisoned for a white-collar crime. She has seen therapists for several years and has also been prescribed antidepressants. The medication has not been very helpful, and in part because of that, she resists an explanation in terms of neurobiological malfunction. But her reasoning is "if it was really something that was just brain chemistry and you took medication, you would be fine." That is "not true" for her, and she doesn't think it is true for others either. The fact that the drugs don't cure mental disorders is for her a sign that something other than brain chemistry is involved.

11. Kramer, *Listening to Prozac*.

12. Slater, *Prozac Diary*, 118–19. In an earlier best seller, psychiatrist Ronald Fieve (*Mood-swing*, 23) gave clear expression to the same idea: "findings point to the fact that

mania and mental depression must be due to biochemical causes handed down through the genes, since they are correctable so rapidly by chemical rather than talking therapy." The reasoning is the inverse of a logic that anthropologists have observed in a completely different context of medication use: "If the problem is physical, then the remedy should be physical" (Whyte et al., *Social Lives of Medicines*, 46). Here, if the remedy is physical, then the problem must be physical as well.

13. Thus, the claim by sociologist David Karp (*Speaking of Sadness*, 77) that medication by itself triggers a transformation in understanding seems overstated. He writes, "The moment individuals first take a drug for depression surely begins a process of biochemical revision."

14. The following quote from a former physician director of mental health services at Harvard University is perhaps relevant to the logic of this position: "I have seen many students over the years who were resistant in principle to medication and had a number of other issues that we worked on in psychotherapy. They eventually decided to try medication because their symptoms persisted, and it was not unusual that after a month or two, the student would say, 'You know this medication has made a huge difference for me, and although I enjoy talking with you about these other issues, I feel like myself again. I'd rather be out with friends and come to see you when needed for refills.' I'm not implying psychotherapy isn't valuable, but biology is sometimes 70 to 80 percent of the problem." Kadison and DiGeronimo, *College of the Overwhelmed*, 249. The distinction Kadison is making between psychological and biological phenomena is common among mental health professionals. A study of clinicians found that they regarded medication to be more effective for disorders they regarded as biologically based, and psychotherapy more effective for disorders they thought had psychosocial origins. Ahn, Proctor, and Flanagan, "Mental Health Clinicians' Beliefs."

15. For further discussion of the aims of psychotherapeutic rationales, see Davis, *Accounts of Innocence*, chapter 5.

16. In the rationale of CBT, symptoms are believed to result from distorted perceptions of reality and reactions based on erroneous assumptions that the individual has internalized. To do CBT is to change the habitual patterns of thought that give rise to the symptoms.

17. Kleinman, *Rethinking Psychiatry*; Frank and Frank, *Persuasion and Healing*.

18. Only one participant indicated that she had never attended psychotherapy because of money. The question of payment did arise for some who had attended therapy, with several reporting disruptions because of changes in insurance coverage or other financial issues. Despite considerable changes in the public and private financing of mental health care, however, the use of psychotherapy, and the sociodemographic characteristics of psychotherapy clients, has remained remarkably stable since at least the late 1980s. Olfson and Marcus, "National Trends in Outpatient Psychotherapy."

19. Kadushin, *Why People Go to Psychiatrists*, 15. Many other studies supported this idea. See, for example, studies and discussion in Garfield, "Research on Client Variables in Psychotherapy."

20. Christopher, a forty-six-year-old white man, who had problems with attention and organization, gave this response to our question about seeing someone: "No. [laughter] I couldn't do it, wouldn't happen, no way." He intimated that there would be some shame in talking with a therapist, a view expressed by an African American man, Dwayne, forty-seven, who was dealing with sadness and isolation.

Although he indicated that sometimes he felt like he would like to seek counseling, he couldn't because there was a stigma on such help-seeking among black men. If he told his friends that he was seeing a therapist, "they would look at me like something's wrong with me," and it seems he would fear as much himself. "I don't want to feel like I'm going crazy," he said.

21. Crossley, "Prozac Nation and the Biochemical Self," 250.

22. This distinction reproduces a contrast between "real" and "psychosomatic" that has a long history in medicine. See, for example, Kirmayer, "Mind and Body as Metaphors."

23. Taylor, *Modern Social Imaginaries*, 23. Taylor speaks of "social imaginaries" as commonly held ways of imagining collective life that enable, by making sense of, the practices of a society. He draws on Benedict Anderson's theorizing of "imagined communities." The notion of a social imaginary, first used by Cornelius Castoriadis in *The Imaginary Institution of Society*, has a strong affinity with Emile Durkheim's much earlier notions of "collective consciousness" and "collective representations" as the glue that holds societies together. For a discussion of these affinities, see Rundell, "Durkheim and the Reflexive Condition of Modernity." Other examples include "technoscientific imaginaries," used by the anthropologist George Marcus ("Introduction," 4) for the ways that scientists imagine future possibilities of innovation in light of practical changes and growing ambiguities in the conditions of scientific work. Linguist Shelia Jasanoff ("Future Imperfect," 4) speaks of "sociotechnical imaginaries," by which she refers to "collectively held, institutionally stabilized, and publicly performed visions of desirable futures, animated by shared understandings of forms of social life and social order attainable through, and supportive of, advances in science and technology."

24. Jones, Kahn, and Macdonald, "Psychiatric Patients' Views"; Weinstein and Brill, "Social Class and Patients' Perceptions of Mental Illness"; Wall and Hayes, "Depressed Clients' Attributions of Responsibility." In another type of study, researchers have explored the causal beliefs of persons who may be undiagnosed but screen positive for depressive symptoms using standard depression scales. These studies too show a growing emphasis on biology and the use of drugs. In one such study of nearly eleven thousand young people (ages sixteen to twenty-nine), some 45 percent of whom had a treatment history, the researchers found "a strong belief in a biological cause and medical treatment approach for depression." Van Voorhees et al., "Beliefs and Attitudes," 42. Also see Goldstein and Rosselli, "Etiological Paradigms of Depression"; and Givens et al., "Ethnicity and Preferences for Depression Treatment." An analysis of many studies of this kind, from both the US and other countries, concluded that "depressed patients believe more in a biological etiology and have more trust in medication than non-depressed patients." Prins et al., "Health Beliefs," 1046.

25. Pescosolido et al., "'A Disease Like Any Other'?" Similarly, a comprehensive review and analysis of studies from many countries found a "coherent trend . . . towards a biological model of mental illness," with the highest rates in the United States. Further, the study's authors note that "conceptualizing mental disorder as a brain disease or a medical problem facilitates the acceptance of a medical-professional solution for this problem." See Schomerus et al., "Evolution of Public Attitudes about Mental Illness," 440, 449.

26. For example, Lebowitz, "Biological Conceptualizations of Mental Disorders."

27. For example, McHugh et al., "Patient Preference for Psychological vs. Pharmacologi-

cal Treatment," which is a meta-analysis of thirty-four studies published through mid-2011. The majority of the reviewed studies concerned depression or anxiety, and many involved participants who had never actually sought help for a mental health issue.

28. Olfson and Marcus, "National Trends in Outpatient Psychotherapy," 1456; Nordal, "Where Has All the Psychotherapy Gone?"

29. Olfson and Marcus, "National Trends in Outpatient Psychotherapy," 1460. The number of those who made only one or two visits a year remained stable from 1987 to 2007 at about one-third of psychotherapy clients (34 percent in 1987, 38 percent in 2007), but there was a decline in those who had more visits, especially in the number of those receiving "longer-term psychotherapy," meaning twenty or more visits in a year. In 1987, 16 percent of therapy clients went that often, but by 2007 the number had fallen to 9 percent.

30. Also see Olfson et al., "National Trends in the Mental Health Care of Children, Adolescents, and Adults."

31. Brief therapy, especially cognitive behavioral, now has a stronghold in training programs. In a systematic study of theoretical orientations represented in clinical psychology graduate training programs, Heatherington and colleagues found doctoral programs at comprehensive universities heavily dominated by a single perspective: cognitive behavioral theory and therapy. See Heatherington et al., "Narrowing of Theoretical Orientations."

32. Levenson and Davidovitz, "Brief Therapy Prevalence and Training," 335.

33. It should be noted that the trend toward briefer therapies predates the shift toward managed care in the US and is similar in countries, like England, which have a very different system of health coverage and do not permit direct-to-consumer advertising of prescription-only drugs. See, for example, Darian Leader. "A Quick Fix for the Soul," *Guardian*, September 8, 2008, http://www.theguardian.com/science/2008/sep/09/psychology.humanbehaviour.

34. Levenson and Davidovitz, "Brief Therapy Prevalence and Training," 337.

35. See, for example, Gottlieb, "The Branding Cure"; and Totton, "Two Ways of Being Helpful."

CHAPTER TWO

1. The concept of healthscape is from Clarke, "From the Rise of Medicine to Biomedicalization."

2. In the first half of the twentieth century, medical breakthroughs—the discovery, for instance, of the etiology of general paresis (an infection of the brain or spinal cord that comes from untreated, late-stage syphilis) and a specific treatment for that condition—offered potential paradigms for a disease model of mental illness. Within a short span in the mid-1930s, a whole series of new physical therapies were introduced for the treatment of schizophrenia. These included insulin therapy, metrazol shock therapy, electroconvulsive therapy (which, by coincidence, was found to be effective with mania and depression), and prefrontal lobotomy. Not only were these treatments widely used during the 1940s and 1950s, they spurred extensive research into their method of action and generated hypotheses about the physiological roots of mental disorders. For a detailed history of somatic treatments in this period, see Braslow, *Mental Ills and Bodily Cures*.

3. Hale, *Rise and Crisis*, 276.

4. Hale; Wilson, "DSM-III and the Transformation of American Psychiatry." David

Healy (*Let Them Eat Prozac*, 4) writes, "Popular thinking was also influenced by the world wars, which brought home the idea that extreme environmental stress could produce 'nervous breakdowns.'"

5. Hale, *Rise and Crisis*, 289.

6. Joint Commission, *Action for Mental Health*, 250, 149, 251.

7. See, for example, the initial US report on the effectiveness of chlorpromazine, where the author emphasizes that chlorpromazine "must be considered a true therapeutic agent with definite indications," not "merely a chemical restraint," but "it should never be given as a substitute for analytically oriented psychotherapy." Winkelman, "Chlorpromazine," 21.

8. Joint Commission, *Action for Mental Health*, 41. See the discussion of the tangled relationship between the new drugs and the deinstitutionalization of the mentally ill in Scull, *Madness in Civilization*, chapter 12.

9. Veroff, Kulka, and Douvan, *Mental Health in America*, 10.

10. While psychoanalysts have often resisted somatic treatments, the relationship is complex and not as dogmatic as it is often portrayed. See, for example, Sadowsky, "Beyond the Metaphor of the Pendulum."

11. Rieff, *Freud*, 354.

12. Rieff, 354, The Freud quote is from his *Three Essays on Sexuality*.

13. DSM-II, 39.

14. Srole et al., *Mental Health*.

15. For example, Hollingshead and Redlich, *Social Class*; Goldhamer and Marshall, *Psychosis and Civilization*; Gurin, Veroff, and Feld, *Americans View*.

16. March and Oppenheimer, "Social Disorder and Diagnostic Order," i38.

17. Rieff, *Freud*, chapter 10.

18. Veroff, Douvan, and Kulka, *Inner American*, 532. See also the companion volume: Veroff, Kulka, and Douvan, *Mental Health in America*.

19. Two histories, which I draw upon here—Tone, *Age of Anxiety*, and Herzberg, *Happy Pills*—situate the conception of anxiety that predominated in the 1950s and 1960s within the "nervous illness" tradition that dates to the late nineteenth century. At that time, many Americans were being diagnosed with "neurasthenia" ("tired nerves") for emotional and psychosomatic problems such as worry, fatigue, muscle pain, and inability to concentrate. Neurasthenia was regarded as a byproduct of the pace and intensity of industrial society, and therefore was most likely to afflict elites, the "brainworkers." Psychiatry at that time was of low status, associated with mental asylums and the treatment of the seriously mentally ill. As sufferers of neurasthenia "did not want their ailments to be confused with psychiatric ones," as Tone writes (14), they took their problems to neurologists, who had higher status, office-based practices, and treatments aimed to help people feel better. In the 1950s, this pattern was repeated. Anxiety was characterized as an aliment of the successful and distinguished from serious mental illness, with treatment sought from regular physicians rather than mental health professionals.

20. These quotes are taken from the following Thorazine advertisements, respectively: *Psychosomatic Medicine* 18, no. 6 (1956); *Mental Hospitals* 7, no. 7 (1956); *Mental Hospitals* 11, no. 5 (1960); and *Psychosomatic Medicine* 18, no. 5 (1956).

21. Pieters and Snelders, "Psychotropic Drug Use."

22. Hollister, "Drugs for Emotional Disorders," 943.

23. Smith, *Social History of the Minor Tranquilizers*, 22.

24. Parry, "Use of Psychotropic Drugs," 801. Before minor tranquilizers came on the

market, ordinary Americans were remarkably open to using the barbiturates. If we look back to the mass popularity of opium and so-called patent medicines (tonics, elixirs, and syrups, often laced with unspecified stimulant and sedative ingredients) in the nineteenth and early twentieth centuries, we might wonder if there was ever a time when people were not receptive to such chemical agents. See Tone, *Age of Anxiety*; and Healy, *The Antidepressant Era*.

25. Herzberg, *Happy Pills*, 34.
26. Herzberg, 34.
27. Ritalin advertisements in *Mental Hospitals* 9, no. 3 (1958), and *Mental Hospitals* 10, no. 2 (1959). One ad, from 1959, cited "a recent study of 9 alcoholic patients," which showed that after Ritalin use, "all patients became significantly more involved in therapy," and "two of the patients . . . for whom intensive individual therapy was available, moved rapidly toward new insights with remarkable emotional participation." *Mental Hospitals* 10, no. 2 (1959).
28. Herzberg, *Happy Pills*, 37.
29. Herzberg, 36–37.
30. Tone, *Age of Anxiety*, 72.
31. Herzberg, *Happy Pills*, 43.
32. Tone, *Age of Anxiety*, 90.
33. Parry, "Use of Psychotropic Drugs," 801.
34. Parry et al., "National Patterns," 774.
35. Blackwell, "Psychotropic Drugs."
36. Tone, *Age of Anxiety*, ix.
37. Smith, *Social History of the Minor Tranquilizers*, 31–32, table 4.1.
38. Parry et al., "National Patterns," 777, table 10.
39. Parry et al.; see also Smith, *Social History of the Minor Tranquilizers*, 31–32, table 4.1.
40. For a particularly clear and influential statement of the insignificance of diagnosis to psychoanalysis, see Menninger, *The Vital Balance*.
41. Blackwell, "Psychotropic Drugs," 1638.
42. Quoted in Marshall, Georgievskava, and Georgievsky, "Social Reactions to Valium and Prozac," 97.
43. Shapiro and Baron, "Prescriptions for Psychotropic Drugs," 486.
44. For example, Blackwell ("Minor Tranquilizers," 29) writes, "Surveys in general practice populations have shown clearly that patients with emotional illness commonly present through the medium of physical complaints."
45. Shapiro and Baron, "Prescriptions for Psychotropic Drugs." See also Herzberg, *Happy Pills*, 38.
46. Blackwell, "Minor Tranquilizers," 29. See also Blackwell, "Psychotropic Drugs," and Parry et al., "National Patterns." According to an analysis of drug prescribing in England published in 1975, "Probably only about 16% of all psychoactive drug takers take the drugs for more specific conditions. The remainder seem to be taking them for vague conditions," and the drugs are "increasingly prescribed to try and modify personal and interpersonal processes." Trethowan, "Pills for Personal Problems," 750.
47. Olfson and Klerman, "Trends in the Prescription of Psychotropic Medications," 561.
48. Cooperstock and Lennard, "Some Social Meanings of Tranquilizer Use." For additional studies from this period, see Lader, "Benzodiazepines"; Manheimer et al., "Popular Attitudes"; Mellinger et al., "Psychic Distress, Life Crisis"; and Waldron, "Increased Prescribing."

49. Tone, *Age of Anxiety*, 27–28.
50. Hale, *Rise and Crisis*, 382.
51. Trethowan, "Pills for Personal Problems," 750.
52. Coppen, "Biochemistry of Affective Disorders," 1237.
53. Hollister, "Psychopharmacology in Its Second Generation," 372.
54. These quotes are from Metzl, "'Mother's Little Helper.'"
55. Coppen, "Biochemistry of Affective Disorders"; Schildkraut and Kety, "Biogenic Amines and Emotion."
56. For an overview of the major importance of depression in research, see Kline, *Factors in Depression*.
57. Hollister, "Psychopharmacology in Its Second Generation," 373.
58. Hollister, 371.
59. Hollister, "Drugs for Emotional Disorders," 944. Previous histories, such as Wilson's influential "DSM-III and the Transformation of American Psychiatry," greatly underestimate the role of drugs in the DSM revolution.
60. Smith, *Social History of the Minor Tranquilizers*, 158.
61. See examples in Smith, chapter 9.
62. Smith, 184. Another sign of the change toward a more medical approach can be seen in front-cover ads in *JAMA* in late 1978, wherein it is announced that Valium "isn't for common tensions of life—tight schedules, traffic jams, final exams. . . . It isn't even for many of the tensions and anxieties patients complain about, for example, those due to normal fears and apprehensions about an illness or diagnostic procedure or surgery." These, of course, were just the sorts of problems that Valium was being marketed for just a few years earlier (cited in Herzberg, *Happy Pills*, 145–46; original emphasis).
63. Hollister, "Drugs for Emotional Disorders," 945.
64. Hollister, 944.
65. Hollister, "Psychopharmacology in Its Second Generation," 372.
66. Snyder et al., "Drugs, Neurotransmitters, and Schizophrenia."
67. Blackwell, "Psychotropic Drugs," 1638.
68. Healy, *The Antidepressant Era*. Also see Rosenberg, *Our Present Complaint*, chapter 3, and Davis, "Reductionist Medicine and Its Cultural Authority."
69. Luhrmann (*Of Two Minds*, 49) notes that a medication response altering a diagnosis, as seen in psychiatry, "is not unlike the rest of medicine. ('Take the antibiotics, and if the rash doesn't go away, we'll know it wasn't Lyme disease.')"
70. For examples in other areas of medicine, see Greene, *Prescribing by Numbers*.
71. Hollister, "Psychopharmacology in Its Second Generation," 373.
72. Hollister, "Drugs for Emotional Disorders," 945.
73. Snyder et al., "Drugs, Neurotransmitters, and Schizophrenia," 1244.
74. In the 1970s, researchers believed that anxiety was regulated by the neurotransmitter GABA (gamma-aminobutyric acid), whose action was facilitated by the benzodiazepines. Gray, "Anxiety and the Brain."
75. A summary article by DSM-III task force chair Robert Spitzer and colleagues, "DSM-III: The Major Achievements," gives some other examples. The definition of schizophrenia, for instance, was narrowed in part because of "differential response to somatic therapies," and the definition of affective disorder enlarged for the same reason. In the new anxiety category, "panic disorder" was differentiated from "generalized anxiety disorder" because of a "differential response" to medication treatment "as compared with other disorders in which anxiety is prominent" (158). On

the relationship between medication response and anxiety categories, see Klein, "Delineation of Two Drug-Responsive Anxiety Syndromes," and Klein, Zitrin, and Woerner, "Antidepressants, Anxiety, Panic, and Phobia." Also see Eysenck, *The Biological Basis of Personality*. For a careful study of the relationship between the clinical effects of drugs and the definition of depression and anxiety disorders, see Healy, *The Antidepressant Era*. The new category "attention deficit disorder," with or without hyperactivity (later, attention deficit/hyperactivity disorder), to give another example, was theorized in terms of cognitive deficits in sustained attention and impulse control, precisely those deficits that stimulant medications were believed to reduce. See Weiss et al., "Studies on the Hyperactive Child"; Douglas, "Stop, Look and Listen."

76. Tourney, "History of Biological Psychiatry," 37. In the early 1970s, Nathan Kline, who was instrumental in the use of iproniazid for depression, observed that the tricyclic antidepressants, such as imipramine, alone had "accounted for some 5,000 scientific papers over the last fifteen years." Kline, *From Sad to Glad*, 131.

77. Developed by neo-somaticists at Washington University in Saint Louis. For more on the Feighner criteria and their development, see Feighner et al., "Diagnostic Criteria"; Feighner "The Advent of the Feighner Criteria."

78. Frances et al., "DSM-IV Classification," 823. Also see Hyman, "Diagnosis of Mental Disorders."

79. The number of diagnostic categories would continue to grow in subsequent editions of the DSM. The DSM system can give no account of the shifting standards of normalcy that underlie this growth and tacitly inform the categories and their symptoms.

80. DSM-II, 39.

81. DSM-III, 228. These criteria would be elaborated considerably in later editions of the DSM.

82. DSM-II, 39.

83. DSM-III, 376. Drug action may have also played some role in this limitation. According to major researcher Donald Klein—a member of three DSM-III advisory committees, including the one on anxiety and dissociative disorders—and colleagues, antianxiety drugs did not diminish conditioned avoidance or compulsive rituals but were "effective only in the presence of overt, manifest anxiety." Klein, Zitrin, and Woerner, "Antidepressants, Anxiety, Panic, and Phobia," 1401; see also Klein, "Delineation of Two Drug-Responsive Anxiety Syndromes."

84. DSM-III, 7.

85. Rosenberg, *Our Present Complaint*, 19.

86. In the DSM-5, published in 2013, the judgment that one's response be excessive or unreasonable is transferred to the clinician, though, again, the basis of that judgment will come from the patient's own report.

87. DSM-III, 7. This explicit claim was dropped with the DSM-IV, though it remains implicit.

88. Hollister, "Drugs for Emotional Disorders," 943.

89. Spitzer, Williams, and Skodol, "DSM-III: The Major Achievements," 152.

90. DSM-III, 7.

91. DSM-III, 6.

92. Pliny Earle, quoted in Frances et al., "DSM-IV Classification," 825.

93. Thomas Insel, NIMH Director's Blog: "Transforming Diagnosis," April 29, 2013, http://www.nimh.nih.gov/about/director/2013/transforming-diagnosis.shtml.

94. In a letter to the APA in November 1974, Robert Spitzer, head of the task force, declared that the DSM-III would be a "defense of the medical model as applied to psychiatric problems." Quoted in Wilson, "DSM-III and the Transformation of American Psychiatry," 405.

95. Wilson, 403. When in 1975 one of the DSM-III task force members wrote to its chair, Robert Spitzer, to observe that very few of the symptom clusters in the proposed manual "could be called diseases" and therefore should just be identified as symptoms, Spitzer sent a memorandum to the task force rejecting this idea not only because it would impede research but because if the manual did not label the symptom clusters "syndromes" or "disorders," it would impede insurance reimbursement. (Wilson, 405.) Already in 1974, the APA Task Force on Nomenclature and Statistics identified serving as a basis for insurance reimbursement as one of the needs they hoped the DSM-III would fill. See Cooper, *Classifying Madness*, 133.

96. For a short history, see Crossley, *Contesting Psychiatry*, chapter 5.

97. Friedan, *Feminine Mystique*; Millet, *Sexual Politics*; Chesler, *Women and Madness*.

98. Rosen, *Psychobabble*.

99. Medawar, "Victims of Psychiatry," quoted in Hale, *Rise and Crisis*, 3.

100. Rosen, *Psychobabble*; Gross, *The Psychological Society*.

101. In the study, Rosenhan ("On Being Sane") and seven confederates presented themselves at hospitals complaining of hearing voices that spoke words that sounded like "'empty,' 'hollow,' and 'thud.'" Besides their real name and occupation, these "pseudopatients" did not change anything else about themselves or their behavior. Each was admitted to the hospital and, with one exception, was diagnosed as schizophrenic. Despite otherwise acting normally and making no further claims to auditory hallucinations, the staff regarded the pseudopatients as disturbed, never caught on that they were being conned, and discharged the pseudopatients, after a period ranging from seven to fifty-two days, with a diagnosis of schizophrenia "in remission."

102. Luhrmann, *Of Two Minds*, 222.

103. Hale, *Rise and Crisis*, 331–32; Wilson, "DSM-III and the Transformation of American Psychiatry," 403.

104. This observation is from philosopher Iris Murdock (*Sovereignty of Good*, 14) in an essay first published in 1964.

105. Hale, *Rise and Crisis*, 301.

106. Herzberg, *Happy Pills*, 157; on the redefinition of depression, see Healy, *The Antidepressant Era*.

107. I put "antidepressant" in quotes because over time the SSRI class has been put to an ever-growing number of clinical uses beyond depression, from the anxiety disorders to eating disorders to chronic pain to fibromyalgia and, in select cases, to much more.

108. Grebb and Carlsson, "Introduction and Considerations," 1.

109. Thomas Insel, NIMH Director's Blog: "Transforming Diagnosis," April 29, 2013, http://www.nimh.nih.gov/about/director/2013/transforming-diagnosis.shtml. For Insel and others of this school, real progress will come from the identification of etiological mechanisms through neuroscience and molecular genetics, which will in turn make possible validated laboratory or imaging biomarkers and specific treatment interventions. One of the large-scale efforts in this regard is the Brain Research through Advancing Innovative Neurotechnologies (BRAIN) Initiative, which was launched in 2013 and is "aimed at revolutionizing our understanding of the human brain." The initiative is led by the National Institutes of Health, the Defense

Advanced Research Projects Agency, the National Science Foundation, and the Food and Drug Administration. On the role of the NIH, see http://www.braininitiative .nih.gov/index.htm.

110. Grebb and Carlsson, "Introduction and Considerations," 1.

111. Ross, Travis, and Arbuckle, "The Future of Psychiatry as Clinical Neuroscience," 413. See also Insel and Quirion, "Psychiatry as a Clinical Neuroscience Discipline," and Reynolds et al., "The Future of Psychiatry as Clinical Neuroscience."

112. All quotes are from Ross, Travis, and Arbuckle, "The Future of Psychiatry as Clinical Neuroscience," 413.

113. See, for example, Hyman, "The Daunting Polygenicity of Mental Illness."

114. Nesse and Stein, "Towards a Genuinely Medical Model," 7.

115. Hyman, "Diagnosis of Mental Disorders," 159.

116. In an analysis of the use of psychiatric medication among over two million insured Americans, Medco Health Solutions ("America's State of Mind"), a prescription drug management company, reported that more than 20 percent of adults (15 percent of men; 26 percent of women) were taking at least one prescription psychotropic in 2010. Report available at http://static1.1.sqspcdn.com/static/f/1072889/ 15159625/1321638910720/Psych+Drug+Us+Epidemic+Medco+rpt+Nov+2011.pdf ?token=OiMneT8RYF6ejLJyMwPx84tmPzs%3D.

117. For minor tranquilizer use in 1969–70, see Parry et al., "National Patterns." For antidepressant use in 2012, see Kantor et al., "Trends in Prescription Drug Use."

118. In the drug use survey for the year 1969–70, the majority of those taking a minor tranquilizer had done so on a short-term basis, meaning less than two months of daily use and intermittently on fewer than thirty-one occasions. Parry et al., "National Patterns." Another national survey, conducted by some of the same researchers in 1979, defined "long-term use" as regular daily use for a year or more. They found that 15 percent of all those taking a tranquilizer were long-term users by that definition. Mellinger et al., "Prevalence and Correlates." By contrast, the most popular drugs now, the antidepressants of the SSRI and SNRI (serotonin and norepinephrine reuptake inhibitors) class are slow onset, typically taking two to four weeks for a patient to experience their effects (if they experience any effects at all; the nonresponse rate is in the range of 40 percent). And in sharp contrast with the earlier period, the great majority take them for long periods of time. In a 2011 study, for instance, covering the years 2005–8, researchers found that among Americans taking antidepressant medication, 30 percent had regular daily use from two months to two years, 30 percent from two to five years, 18 percent from five to ten years, and an additional 14 percent for over ten years. Pratt et al., "Antidepressant Use in Persons Aged 12 and Over," 4, figure 4.

119. Frye et al., "The Increasing Use of Polypharmacotherapy," 9. For a review of studies, which showed a "significant decline" in the rate of monotherapy between the 1970s and the 1990s, see Rittmannsberger, "The Use of Drug Monotherapy." Also see Kantor et al., "Trends in Prescription Drug Use."

120. Gardos et al., "Polypharmacy Revisited," 179.

121. Baum et al., "Prescription Drug Use in 1984 and Changes over Time," 112.

122. Howie, Pastor, and Lukacs, "Use of Medication for Emotional or Behavioral Difficulties," 1.

123. Mojtabai and Olfson, "National Trends in Psychotherapy."

124. Barber, *Comfortably Numb*, 199. See also, for example, Kandel, "A New Intellectual Framework for Psychiatry," and Vaughan, *The Talking Cure*.

125. In 2000, for example, psychoanalysts formed an International Neuropsychoanalysis Society, which hosts a yearly conference for researchers, clinicians, and students and publishes the journal *Neuropsychoanalysis*. The journal publishes "papers at the intersection of psychoanalysis and the neurosciences" and was "founded on the assumption that these two historically divided disciplines are ultimately pursuing the same task."

126. Insel and Quirion, "Psychiatry as a Clinical Neuroscience Discipline," 2223. Reynolds and colleagues ("The Future of Psychiatry as Clinical Neuroscience," 446) offer these examples of the tools that psychiatry has available for developing new assessment and treatment approaches: "brain imaging, genetics, neuropsychopharmacology, neurophysiology, epidemiological models of risk and protective factors, and neuropsychology."

CHAPTER THREE

1. DSM-III, xxii.

2. When the interview began, we asked what problem led the participant to respond to our notices. We already knew the short answer since people had called an 800-number in order to prequalify for the study and they had been asked questions about the problem, if they had seen a medical professional, whether they were taking any medications and for how long, and so on. But we did not use this specific knowledge in the interview.

3. I will consider how people who took on a medicalizing perspective relate to the question of science in chapter 6.

4. People engaged medical concepts with their own "active, responsive attitude," to use a phrase from philosopher of language Mikhail Bakhtin, who is famous for his emphasis on language use as a "living dialogue." Bakhtin argues that in "live speech communication," as we deal with particular situations in light of our particular intentions and values, we take words from others and, more or less creatively, we "assimilate, rework, and re-accentuate" them in order to make them our own. These assimilated or appropriated words and phrases, which always bear the echoes of previous uses, are thus imbued with our expression and reflect our evaluative stance. Adding a new layer of meaning, others' words become our words. The medical concepts that participants encountered and made their own had this appropriated character. Bakhtin, *Speech Genres*, 68, xv, 88, 89.

5. Stromberg, "The Impression Point," 61. This conceptualization builds on the theorizing of Wilhelm Dilthey in the hermeneutic tradition.

6. For a wider range of examples of such dialectical self-transformations, see DeGloma, *Seeing the Light*.

7. In a study of popular depression memoirs, sociologist Christina Simko makes a similar point about the need to contend with a dominant psychiatric language. She writes, "Yet ironically, these memoirs . . . illustrate the cultural power of the biomedical account. It cannot be ignored, but—if it is not adopted—it must be grappled with, overtly rejected, and even polemicized against. In short, as [Anne Hunsaker] Hawkins argues, oppositional narratives illustrate the durability and salience of the biomedical model: 'That this myth can generate complaints of its ineffectiveness or unhelpfulness is actually a sign of its vitality. The mark of a living myth is not so much its veracity, nor even its utility . . . but its authority and power, the degree to which people feel compelled to believe in it.' [Hawkins, *Reconstructing Illness*, 33–

34] Rejecting it requires an explicit justification." Simko, "The Problem of Suffering in the Age of Prozac," 76.

8. For a review of studies showing the active role of patients—drawing on their own understandings, experiences, and interaction with doctors—in the process of medication prescribing and use, see Malpass et al., "'Medication Career' or 'Moral Career'?"

9. Besides seeking patient compliance with the recommended treatment, another reason for referencing neurochemistry is likely psychotherapeutic. Telling the patient that the drug fixes a neurobiological imbalance or misfiring may prime the patient to view the medication more optimistically and thus enhance the actual effect of the drug. Physicians are fully aware of the placebo phenomenon that is inherent in routine clinical care and actively work to promote these therapeutic effects.

10. The gap I found between clinical and everyday understandings and practices is not unusual. Research studies have demonstrated such a gap in a variety of different contexts. Studies of doctor-patient encounters and lay perspectives, for example, show varying degrees of distance between lay understandings and medical discourses. See, for example, Karp, *Speaking of Sadness*; Lupton, *Medicine as Culture*; Williams and Calnan, "The 'Limits' of Medicalization?"

11. Among many studies exploring appropriation, see Aikin, Swasy, and Braman, "Patient and Physician Attitudes and Behaviors." See, further, Nichter, "The Mission within the Madness," and Bury, *Health and Illness in a Changing Society*. Bury reviews studies that show how patients seek to influence the outcome of doctor visits by selecting and ordering the information they provide. For evidence of the casual diagnosing of physicians, see, for example, Zimmerman and Galione, "Psychiatrists' and Nonpsychiatrist Physicians' Reported Use of the DSM-IV Criteria"; and Carlat, *Unhinged*.

12. Rose, *The Politics of Life Itself*, 110. Also see Armstrong, "The Patient's View"; Greco, "Psychosomatic Subjects and the 'Duty to Be Well'"; and Mol, *The Logic of Care*.

13. Mayes and Horwitz, "DSM-III and the Revolution in the Classification of Mental Illness," 250.

14. The DSM recognizes that some problems may be brought to "clinical attention" but for which "the individual has no mental disorder." DSM-IV-TR, 731.

15. Healy, *Let Them Eat Prozac*, 6.

16. Rose and Novas, "Biological Citizenship."

17. Ross, *Triumph over Fear*; Stein and Walker, *Triumph over Shyness*.

18. Schneier and Welkowitz, *The Hidden Face of Shyness*; Bourne, *The Anxiety and Phobia Workbook*.

19. Slater, *Prozac Diary*; Stossel, *My Age of Anxiety*; Solomon, *The Noonday Demon*. On pathographies, see Davis, "Post-Prozac Pathography"; Hawkins, *Reconstructing Illness*; and Simko, "The Problem of Suffering in the Age of Prozac."

20. Clarke and Gawley, "The Triumph of Pharmaceuticals," 95.

21. Clarke and Gawley, 97, 99.

22. Metzl and Angel, "Assessing the Impact of SSRI Antidepressants."

23. These participant observations are consistent with the negative critiques of DTCA offered by social scientists and ethicists. Critics typically charge this advertising with hyping drugs to people who do not actually need them, creating unrealistic expectations of effectiveness, glossing over side effects, driving up the cost of prescription drugs, and harming the doctor-patient relationship. See, for example, Hoffman and Wilkes, "Direct to Consumer Advertising."

24. Ads not only "say it," they may also offer visual simulations of chemical imbalance, and several participants commented on "dramatizations" they had seen at some point in commercials or on drug websites.

25. Smith, *Social History of the Minor Tranquilizers*, 184.

26. Cohen, "Direct-to-the-Public Advertisement of Prescription Drugs," 373. By 1997, the year the FDA eased the DTCA rules, nearly eighty pharmaceuticals were being advertised directly to the public. Siegel, "DTC Advertising: Bane . . . or Blessing?" 146.

27. See DTCA expenditure information in Ventola, "Direct-to-Consumer Pharmaceutical Advertising," 670; and Palumbo and Mullins, "The Development of Direct-to-Consumer Prescription Drug Advertising Regulation," 423. The FDA's initial reservations about DTCA were shared by others. In the early 1980s, all the major medical associations were opposed to DTCA (Cohen, "Direct-to-the-Public Advertisement of Prescription Drugs," 373). The American Medical Association, for instance, imposed a ban on DTCA in its patient publications in 1984. Then, in 1992, it reversed its position with little dissent and lifted the ban. Even the big drug companies were initially opposed to DTCA. See Critser, *Generation Rx*.

28. See Sullivan et al., "Prescription Drug Promotion from 2001–2014," and Mackey, Cuomo, and Liang, "The Rise of Digital Direct-to-Consumer Advertising?" Exposure to social media ads is hard to study because the ads that individual people see are highly targeted based on information gathered about them.

29. Aikin, Swasy, and Braman, "Patient and Physician Attitudes and Behaviors," 2.

30. Aikin, Swasy, and Braman, 2.

31. On patient prescription drug spending, see Dave, "Effects of Pharmaceutical Promotion." Critical discussions of DTCA often frame it as a type of "disease mongering," as an effort to convince "well people that they are sick, or slightly sick people that they are very ill" (Payer, *Disease-Mongers*, 5). This perspective follows a long tradition of treating advertising in terms of indoctrination and manipulation and consumers as passive and easily duped. In what follows, my goal is to consider how DTCA facilitates appropriation by exploring what it offers to people, how it speaks to their agency, desires, dissatisfactions, and hopes. There is persuasion, to be sure, but in my view, it works *with* people not against them, offering a form of self-help in the classical sense, namely, that a better life is available and you have the power to change. See, along these lines, Miller and Rose, "Mobilizing the Consumer." They stress, in their study of postwar advertising, that advertising was "not so much the invention and imposition of 'false needs,' but a delicate process of identification of the 'real needs' of consumers, of affiliating these needs with particular products, and in turn of linking these with the habits of their utilization" (6).

32. See, for example, Davis, "Suffering, Pharmaceutical Advertising, and the Face of Mental Illness"; Woloshin et al., "Direct-to-Consumer Advertisements for Prescription Drugs"; Dumit, *Drugs for Life*, chapter 2; Grow, Park, and Han, "'Your Life Is Waiting!'"; Metzl, *Prozac on the Couch*.

33. Interestingly, they are again text heavy but for a different reason. They don't have to expend a lot of words on explaining a condition, but because of medical and legal problems, especially with the SSRI and SNRI class of medications, descriptions of side effects are far more detailed than previously.

34. See, for example, Healy, *The Anti-Depressant Era*, chapter 6; Lane, *Shyness*.

35. The FDA recognizes and regulates three forms of DTC advertisements: illness-awareness advertisements; reminder advertisements, which simply call attention to

a drug brand; and "product-claim" or "indication" advertisements. The latter name a specific drug and the condition it is approved to treat (its "indication"), along with side effects and contraindications (the "brief summary"). See Palumbo and Mullins, "The Development of Direct-to-Consumer Prescription Drug Advertising Regulation," 427–29.

36. See Cottle, "Selling Shyness." The "Imagine Being Allergic to People" poster is in the author's collection.

37. Conrad and Potter, "From Hyperactive Children to ADHD Adults," 562–65.

38. Note, once again, a revised label; "major depressive disorder" is in the DSM, not "clinical depression." This advertisement is reproduced in Metzl, *Prozac on the Couch*, 161.

39. See, for example, Payton and Thoits, "Medicalization, Direct-to-Consumer Advertising, and Mental Illness Stigma."

40. The listing of *bodily* side effects strengthens the idea that the medication is fighting a real *bodily* condition. According to the chief executive officer of Cult Health, an ad agency that specializes in health care: "It's counterintuitive, but everything in our research suggests that hearing about the risks increases consumers' belief in the advertising." Kaufman, "Think You're Seeing More Drug Ads on TV?"

41. The approach I am taking here is shaped by semiotics, the study of sign systems. In this approach analysis focuses on the ad as a symbolic whole and its connotative levels of communication. For classic works on advertising, see Williamson, *Decoding Advertisements*, and Ewen, *All Consuming Images*.

42. The ad, for instance, appeared in *Parade*, December 20, 2015, 11–12.

43. Dumit, *Drugs for Life*, 58–63.

44. Grow, Park, and Han, "'Your Life Is Waiting!'" Also see Rose, *The Politics of Life Itself*, 25–27.

45. For an illuminating discussion of this point, see Stepnisky, "Narrative Magic."

46. For an advertisement to create meaning, we, the audience, must participate. We need to understand the referents and complete the unspecified but necessary exchange of signifiers. The exchange, when made, is a product of our mind and not simply a perception of what was written or seen. We create the meaning of the ad through our "decoding," an imaginative engagement that gives advertisements their dynamism and invites viewer pleasure and identification. Williamson, *Decoding Advertisements*.

47. The prescription drug ads in medical journals are often aligned with complementary DTCA campaigns or depict categories of sufferers in terms very much like DTCA. In these ads, the images are directed to recognizing common forms of distress in others and imagining a better future for them.

48. Although there are patient forums online and in social media, and these are important for some, only one person mentioned such a forum, and she indicated that she had at one time read some of the stories she found at an ADHD site for adults but had never herself participated.

49. This includes eight participants who had first seen the professional as a child, when parents were obviously involved.

50. See the studies, for example, in Horwitz, *The Social Control of Mental Illness*, chapter 3.

51. On the role of teams in self work, see Goffman, *Presentation of Self in Everyday Life*, chapter 2.

52. While professionals may not control what meaning is given to the clinical categories, they do control the technical procedures of intervention. Greco, "Thinking beyond Polemics."

53. Goffman called such management "face-work," the effort to present a consistent image that reflects approved social attributes and sustains a desired evaluation from others and of oneself. See Goffman's classic discussion "On Face-Work" in *Interaction Ritual*, 5–45. Of course, this effort to contain meaning is taking place in an interview and is in part directed to the impression made on the interviewer. There is no escaping that dynamic, though we did our best not to express evaluations or signal social desirability. However, as the stories of participants will show, they were conscious and intentional in their control over the language in which they framed their experience and identified decision moments as potentially dangerous.

CHAPTER FOUR

1. On the sick role, see Parsons, *The Social System*.
2. Brownlee, "Mysteries of the Mind."
3. Goffman, *Stigma*, 3.
4. Luhrmann, *Of Two Minds*, 8.
5. David Karp, *Is It Me or My Meds?*, 223.
6. Douglas, *Purity and Danger*.
7. Antistigma efforts follow psychiatric definitions of mental illness and so include much of what I am calling everyday suffering.
8. Cumming and Cumming, "Mental Health Education," 96.
9. Cumming and Cumming, *Closed Ranks*, 88.
10. Cumming and Cumming, 20.
11. Quotes in the balance of this paragraph are from Cumming and Cumming, "Mental Health Education," 109, 99, 109, 110.
12. Cumming and Cumming, 114.
13. Quotes in the balance of this paragraph are from Cumming and Cumming, *Closed Ranks*, 101, 21, 137, 136.
14. In 1961, the Joint Commission on Mental Illness and Health, in its report *Action for Mental Health* (discussed in chapter 2), made the same point. They lamented how the public had turned "deaf ears" to the psychiatric "cardinal tenet" that the "mentally ill . . . are sick in the same sense as the physically sick" (59). This tenet, the commission also affirmed, was not only a scientific truth, but critical to overcoming negative attitudes and fostering "humane, healing care" (56).
15. See, for instance, Itkowitz, "Unwell and Unashamed."
16. The quote is from Sarbin and Mancuso, "Failure of a Moral Enterprise," 159.
17. See the "About Mental Illness" webpage at http://www2.nami.org/Content/NavigationMenu/Inform_Yourself/About_Mental_Illness/About_Mental_Illness.htm.
18. See the webpage at https://bringchange2mind.org/learn/.
19. US Department of Health and Human Services, *Mental Health*, 9. While some have argued that terms such as "illness" and "disease" as used in these contexts do not necessarily imply biological causes, the logic of how the "cardinal tenet" ("disease just like another other") leads to humane attitudes is based precisely on that implication. See Read, "Why Promoting Biological Ideology Increases Prejudice," 118.
20. Rüsch et al., "Biogenetic Models of Psychopathology," 328.
21. See the webpage at https://bringchange2mind.org/learn/.
22. Schomerus et al., "Evolution of Public Attitudes about Mental Illness," 441.
23. D'Arcy and Brockman, "Changing Public Recognition of Psychiatric Symptoms?"

24. Goldman, "Progress in the Elimination of the Stigma of Mental Illness."
25. This quote is from Rahav, "Public Images." Cited in Read, "Why Promoting Biological Ideology Increases Prejudice," 119.
26. On professional views, see Stuart and Arboleda-Florez, "Community Attitudes toward People with Schizophrenia"; Gray, "Stigma in Psychiatry"; and Read, "Why Promoting Biological Ideology Increases Prejudice," 123. A time-series study of the effects of DTCA on stigma over a ten-year period found that "respondents increasingly supported talking to doctors and taking prescription drugs for symptoms of major depression, evidence of medicalization in their thinking about this disorder." However, this openness to medicalization had no effect on "stigmatized views of persons with schizophrenia or depression." The study authors conclude that "these preliminary results therefore lend little evidence to support the popular assumption that medicalization reduces stigma." Payton and Thoits, "Medicalization, Direct-to-Consumer Advertising, and Mental Illness Stigma," 62, 55, 62.
27. The pharmaceutical industry, not surprisingly, supports many of these campaigns and the patient advocacy groups that sponsor them. See, for example, Moynihan and Cassels, *Selling Sickness.*
28. As shown, for example, in Goldstein and Rosselli, "Etiological Paradigms of Depression"; Lebowitz, "Biological Conceptualizations of Mental Disorders among Affected Individuals"; Lebowitz, Pyun, and Ahn, "Biological Explanations of Generalized Anxiety Disorder"; Pescosolido et al., "'A Disease Like Any Other'?"; Read, "Why Promoting Biological Ideology Increases Prejudice"; Read and Dillon, *Models of Madness.* The surgeon general report of 1999 noted that "stigma was expected to abate with increased knowledge of mental illness, but just the opposite occurred: stigma in some ways intensified over the past 40 years even though understanding improved" (US Department of Health and Human Services, *Mental Health,* 8).
29. In the view of antistigma advocates, lay ignorance remains the problem: people have finally caught on to the brain-disease idea, but are now wrongly drawing "essentialist" conclusions from it. In this interpretation, the lay public has interpreted brain disease to mean that those afflicted have an unalterable flaw in their biological makeup that causes their unusual emotions and behavior and has mistakenly regarded this flaw as representing a categorical difference between persons with and without disease. From these errors, people then draw a variety of unwarranted and discrediting conclusions about the "essential" (fixed) identity, "otherness," and uncontrollability of the afflicted and express a desire to maintain social distance. See, for example, Lebowitz, Pyun, and Ahn, "Biological Explanations of Generalized Anxiety Disorder."
30. See, for instance, Mehta and Farina, "Is Being 'Sick' Really Better?"
31. This trade-off is nicely captured by a man taking medication who was interviewed by the sociologist David Karp: "You talk about personal identity. The good news is that it's biogenic [and] therefore it's not my fault. The bad news is it's biogenic because I'm just a passenger on life's way, and I have no idea of who's driving me where and to what destination." Karp, *Is It Me or My Meds?,* 104. It is noteworthy that in surveys of the general public and of patients, psychosocial representations of mental distress, which do not in any simple way discard responsibility, are, unlike disease representations, correlated with *decreased* stigma. See, for example, Mehta and Farina, "Is Being 'Sick' Really Better?," and the studies listed by Read, "Why Promoting Biological Ideology Increases Prejudice," 123.

32. National surveys of the general public find that "bad character" and immorality are some of the *least* endorsed causes of mental health problems. See, for example, Link et al., "Public Conceptions of Mental Illness."

33. Goffman, *Stigma*, 128.

34. In fact, I think it is fair to say that personal fault was more rhetorically salient for those who adopted a medicalizing account. This salience may reflect, paradoxically, the central role of fault in the larger antistigma and medical discourse, which makes a central point of insisting that what a medicalized understanding means is that the sufferer is not responsible. In other words, professional rhetoric may create or heighten a concern with personal responsibility that "folk" psychosocial explanations do not. On a similar point, see Schneider and Conrad ("In the Closet with Illness"), who argued that stigma could have its origins in the "stigma coaching" of well-intentioned others.

35. A study of patients with "mood instability" found that "some participants expressed shock and/or fear on receipt of a formal diagnosis, but for many, diagnosis was helpful and contributed to a meaningful explanation of their symptoms. Many participants also felt the receipt of a diagnosis absolved them from feeling excessively responsible for their problems." Bilderbeck et al., "Psychiatric Assessment of Mood Instability," 237. See also Schreiber and Hartrick, "Keeping It Together."

36. Haslam, "Folk Psychiatry," 622.

37. The quote is from Coulter, "Beliefs and Practical Understanding," 165. In context, Coulter is talking not about the medical model but about a dominant strain of sociology in which lay beliefs are generally regarded as false.

38. Cassell, *The Nature of Suffering*.

39. Only one or perhaps two participants suggested that the act of concealment itself caused them distress.

40. In an often-cited study, the sociologists Graham Scambler and Anthony Hopkins found—as have other studies before and since—that among the persons suffering from epilepsy whom they interviewed, only a minority could ever recall an experience of stigma or discrimination directed against them. This led the study authors to make a distinction between "enacted stigma," that is instances of direct discrimination by others on the "grounds of their perceived unacceptability or inferiority," and "felt stigma," which they defined as a fear of enacted stigma and a feeling of shame associated with the condition. Scambler and Hopkins, "Being Epileptic: Coming to Terms with Stigma," 33. Their distinction builds on an earlier distinction in the experience of stigmatized individuals between being "discredited" (the "undesired differentness" is known about or plainly evident) and "discreditable" (the undesired differentness is not known or readily apparent but would be potentially discrediting if revealed) elaborated by Erving Goffman in *Stigma*.

41. As mentioned in chapter 1, a number of African American participants referred to a general stigmatization of mental illness within the African American community. Ella, for instance, thirty and working in retail sales and as a jazz singer, felt that among African Americans, depression was either "something only white folks get" or a sign that "you ain't living right." She has shared her depression diagnosis with only a few people on what she calls a "need-to-know basis."

42. In his famous book, *How to Do Things with Words*, the philosopher J. L. Austin defined a category of speech he called "performative." Unlike utterances that describe or report or simply say something, Austin argued, a performative utterance is one that makes itself true by being said. It is a certain kind of action. He offered such

examples as the "I do" spoken in the context of a wedding and the pronouncing of a ship's name at launch, while smashing a bottle against its bows. These words "perform" the commitment or naming; they bring it into existence. Many participants seemed to worry that acknowledging weakness or failure had this performative character.

43. Karp, *Speaking of Sadness*, 72, 73, 31.

44. The DSM itself does not, strictly speaking, limit a mental disorder to a biological or neurobiological dysfunction, (DSM-IV-TR, xxxi) but, as we saw, in much public discourse, such as the antistigma campaigns, there is a presumed somatic cause. Most participants saw it that way.

45. Like diabetes, suggesting another flaw with that analogy.

46. This included the small number of those who appropriated a category but had not taken medication, either because they had never been formally diagnosed or had been formally diagnosed by a psychologist and only attended psychotherapy sessions.

47. On the use of stimulants without a prescription, see, for example, DeSantis and Hane, "'Adderall Is Definitely Not a Drug'"; Garnier-Dykstra et al., "Nonmedical Use of Prescription Stimulants"; McCabe et al., "Trends in Medical Use, Diversion, and Nonmedical Use."

48. A few participants began using a medication prior to a prescription but then subsequently got a prescription. At the time of the interview, only one participant was regularly using a medication without a prescription. Kyle is a college senior who has taken a stimulant, obtained from friends, at times of assignment due dates and exams through most of college. He believes that he has undiagnosed ADD and so does not see his use as conferring any unfair advantage.

CHAPTER FIVE

1. Following Goffman and others, I am using the concept of norms here more broadly than in the standard sociological meaning of guides to identifiable courses of action.

2. According to Georges Canguilhem (*On the Normal and the Pathological*, 149), the normal and the abnormal or norm-nonconforming are always a joined pair, constituted in relation to each other. However, although the "abnormal comes after the definition of the normal, it is its logical negation," in fact, "existentially" the abnormal comes first. In a historical context, Nikolas Rose (*Inventing Our Selves*, 26) makes a similar point: "our vocabularies and techniques of the person, by and large, have not emerged in a field of reflection on the normal individual, the normal character, the normal personality, the normal intelligence, but rather, the very notion of normality has emerged out of a concern with types of conduct, thought, expression deemed troublesome or dangerous." Also see Hacking, *The Taming of Chance*, on the historical invention of normalcy.

3. We can leave aside the question here of whether some social norms are valued because of our nature, of what we are like as human persons, or are valued because they just happen to be the customs of our particular society. For a careful explication of the nature/culture tension, see Sayer, *Why Things Matter to People*.

4. Archer, *Being Human*, 218.

5. Archer, 217.

6. In a general sense, this ambiguity is a common feature of social life, reflecting the tension between what people reflexively know—of the logic of their actions and

beliefs; the ways they are formed by their circumstances—and what may be inaccessible to them. See Giddens, *Central Problems in Social Theory*, 144.

7. Given that social emotionality arises at the confluence of personal definitions of self-worth and normative order, it follows (or so I suggest) that, excepting the delusional, identifying what people are emotional about can illuminate not only their personal sense of the good but point to the operative social norms as well. I work out this argument at greater length in Davis, "Emotions as Commentaries on Cultural Norms."

8. Douglas, *Purity and Danger*, 109.

9. Of course, we may be unsure about what the good for us is, and we may misapply norms to ourselves, making evaluations and comparisons that are wrong or unwarranted. Our feelings may be at odds with our understanding of a situation, and we may not be able to name emotions we are experiencing. The stories of participants like Rob certainly raise these possibilities. Nothing I write here is meant to suggest otherwise.

10. Norms of being may constitute, to quote Goffman (*Stigma*, 128), "one sense in which one can speak of a common value system in America." He also adds this important observation: "The general identity-values of a society may be fully entrenched nowhere, and yet they can cast some kind of shadow on the encounters encountered everywhere in daily living."

11. We see our environment as a neutral space, "within which," to quote Charles Taylor (*Human Agency and Language*, 4), "we can effect the purposes which we determine out of ourselves." Iris Murdoch (*Sovereignty of Good*, 42) writes, "If the will is to be totally free the world it moves in must be devoid of normative characteristics, so that morality can reside entirely in the pointer of pure choice."

12. Murdoch, *Sovereignty of Good*, 8.

13. Foucault, "On the Genealogy of Ethics," 110–11.

14. Giddens, *Modernity and Self-Identity*, 88–98.

15. The historian Peter Stearns theorizes that the rise of formal support groups starting in the 1950s was partly a way to share (with strangers) emotions that could no longer be shared with friends or family. The same could be said of individual therapy: "The rise of therapy in the American middle class," Stearns writes, "though deriving from a number of factors . . . owed much to the need to find listeners as the more traditional pool dried up" (*American Cool*, 250).

16. See Hochschild, "Emotion Work, Feeling Rules, and Social Structure."

17. The sociologist Eva Illouz (*Saving the Modern Soul*, chapters 3 and 6) shows that before bursting into popular culture, the basic ideas of emotional competence had been a feature of workplace management and other contexts of problem-solving since the 1920s.

18. Goleman, *Emotional Intelligence*, 45.

19. Berlin, *Four Essays on Liberty*, 122–31.

20. For a discussion of a point like this in liberal political theory, see Sandel, *Liberalism and the Limits of Justice*, 59–65.

21. Johnson, *Who Moved My Cheese?* The sales figures are from Sandomir, "Spencer Johnson, Author of Pithy Best-Sellers," D6.

22. Another participant, Phil, fifty, offers a concise statement of this view of autonomy: He says, "I don't think you have to live up to any particular set of ideals, that's the whole point. I think you can define your own ideals. So, I don't think there's any great social pressure to do this, that, or the other thing."

23. I've added "for yourself" here to distinguish the autonomy norm from other conceptions of right order that also include an imperative to improve yourself, but where the purpose is to bind oneself more securely to a scheme of things (God's will, social solidarity, etc.) that is prior to and independent of self.

24. The notion of scale here is an adaptation of Luc Boltanski's ranking scale of people in terms of "size" and "worth," or "greatness and smallness." These qualities do not inhere in persons but are produced in other ways. See Boltanski, *Love and Justice as Competences*, chapter 1; and *Distant Suffering*.

25. As noted in chapter 4, those who did *not* come under medical management were more likely to accept than resist their limitations.

26. See DeSantis and Hane, "'Adderall Is Definitely Not a Drug,'" on college students' distinctions between good and bad drugs; and Vrecko, "Just How Cognitive Is 'Cognitive Enhancement'?," on the role of emotions in the evaluation of stimulant medication effects.

27. Rogers, *On Becoming a Person*, 35, quoted in Illouz, *Saving the Modern Soul*, 159. The reference to Maslow is also from Illouz, 160.

28. These developments represent a shift from earlier positive thinking, such as Norman Vincent Peale's famous book, first published in 1952, *The Power of Positive Thinking*. The notion of self-optimization in Peale is not rooted in the self but in a dependence on God, an affirmation that "the kingdom of God is within you" (17).

29. See, for example, du Gay, "Against 'Enterprise.'" On the notion of persons modeling themselves on the concept of a "brand," see Davis, "The Commodification of Self."

30. Rose, *Inventing Our Selves*, 154. See, further, among many sources, Sennett, *The Culture of the New Capitalism*; and du Gay, "Against 'Enterprise.'"

31. Rosa, *Social Acceleration*, 181.

32. Rosa, 182, emphasis removed. He observes that in light of the driving force of these cultural patterns of meaning and subjective action orientations, "the capitalist organization of the economy does not appear as a *cause* of the ideology of acceleration but rather as its *instrument*" (183, original emphasis).

33. Greco, "Psychosomatic Subjects and the 'Duty to Be Well.'" Writing in the *Washington Post*, a business psychologist (LaBier, "You've Gotta Think Like Google") depicts the connection between the self and enterprise norms. If the internet company Google "were a person," he wrote, "it would be the model of a psychologically healthy adult." Its "corporate culture and management practices," he continued, "depend upon cooperation, collaboration, non-defensiveness, informality, a creative mind-set, flexibility and nimbleness, all aimed at competing aggressively for clear goals within a constantly changing environment."

34. The title of a book by French essayist Pascal Bruckner gets at this seeming illogic: *Perpetual Euphoria: On the Duty to Be Happy*.

35. In his *Listening to Prozac* (17), Peter Kramer draws the connection between norms of drive and optimism and the enterprising self. He observes a way of being that psychiatrists call "hyperthymia." What sets "hyperthymics" apart is that they are "optimistic, decisive, quick of thought, charismatic, energetic, and confident," qualities he notes that "can be an asset in business."

36. US Department of Health and Human Services, *Mental Health*, 5.

37. See, further, Cruikshank, *The Will to Empower*, chapter 4.

38. Schiraldi, *The Self-Esteem Workbook*.

39. See, for example, Hoffman, "Raising the Awesome Child."

40. For a brilliant discussion of flattery in all types of representational media, see de Zengotita, *Mediated*.

41. See, for example, Rogers, *On Becoming a Person*; Rieff, *Triumph of the Therapeutic*; and Illouz, *Saving the Modern Soul*.

42. Giddens, "Living in a Post-Traditional Society," 75; Rose, *Inventing Our Selves*, 100, 17.

43. See, for instance, the discussion in Trilling, *Sincerity and Authenticity*; Taylor, *Sources of the Self*; Taylor, *The Ethics of Authenticity*; Elias, *The Civilizing Process*; and Rieff, *Triumph of the Therapeutic*.

44. Philosopher Charles Guignon suggests a different sense of "authenticity" in contemporary society: "For our modern way of thinking, then, one does not turn inward in order to reach something greater than or outside oneself. On the contrary, one turns inward because it is within the innermost self that one discovers the ordinarily unseen and untapped resources of meaning and purpose." Guignon, *On Being Authentic*, 82–83.

45. Bell, *Cultural Contradictions of Capitalism*, 19.

46. Guignon, *On Being Authentic*, 6.

47. Charles Taylor, writing in this older vein, argues that "authenticity is not the enemy of demands that emanate from beyond the self; it supposes such demands." His list of possibilities includes "the demands of nature, or the needs of my fellow human beings, or the duties of citizenship, or the call of God, or something else of this order" that matters crucially to a person. An authenticity built only on "what I find in myself," apart from a defining "horizon of significance," would be "trivial." Taylor, *Ethics of Authenticity*, 41, 40.

48. Spaemann, *Persons*, 93.

CHAPTER SIX

1. Guignon, *On Being Authentic*, 82.

2. For many of those with a psychologizing perspective, as we have seen, the emphasis was on struggle. While most did emphasize "getting control" (and control signified independence), they expressed improvement—for the time being—in a language of learning and coping. But they too rarely spoke of the need to explore past personal experience.

3. There were a few participants with a medicalizing perspective who did not take a medication. I am only considering here those who did.

4. Yet another difference between participant experience and a treatment for a disease like diabetes.

5. See, for example, the studies cited in Deferio et al., "Using Electronic Health Records."

6. This practice is an instance of what I earlier referred to as the "target-symptom approach" or what epidemiologist Julie Magno Zito ("Pharmacoepidemiology," 966) refers to as "symptom-specific treatment," where the focus is not on the "specificity of drugs for major diagnoses" but on "symptom suppression." This style of treatment is especially evident in "concomitant therapy," which involves multiple medications (polypharmacy) and off-label prescribing, an approach most commonly used with children being treated with drugs for mental health problems.

7. See, for instance, Lennard et al., *Mystification and Drug Misuse*; DeGrandpre, *The Cult of Pharmacology*, which includes discussion of placebo response studies even with powerful and fast-acting drugs like cocaine; and Howard Becker's famous 1953

article, "Becoming a Marihuana User," which outlines the learning process in marijuana use.

8.  Furukawa et al., "Placebo Response Rates in Antidepressant Trials." They defined placebo response as a 50 percent or greater reduction in depression severity score compared to the baseline.

9.  See, for example, Finniss et al., "Biological, Clinical, and Ethical Advances of Placebo Effects."

10. An interesting 2009 review of research studies sought to evaluate whether placebo effects influenced perceptions of behavior or cognition among elementary school–age children being treated with medication for ADHD. They found little evidence for such effects. "However," the study authors observed, "there may be significant placebo effects in adults who evaluate children with ADHD. Evidence suggests that parents and teachers tend to evaluate children with ADHD more positively when they believe the child has been administered stimulant medication and they tend to attribute positive changes to medication even when medication has not actually been administered." Waschbusch et al., "Are There Placebo Effects in the Medication Treatment of Children?"

11. For a detailed discussion of the contemporary criticism of the antidepressants and a rejoinder, see Kramer, *Ordinarily Well.*

12. On this point, see the discussion of objects in Becker, "History, Culture and Subjective Experience."

13. Turner, *The Ritual Process,* 52.

14. Two participants with a medicalizing perspective and taking medication reported little benefit. They, however, like Rob in chapter 5, who complained of a waning drug effect, did not have a change of mind.

15. One might expect that being told of a neurobiological problem, especially in the age of the internet, would lead patients to make it their business to research and understand the medical science behind their problem. Participants could give but the vaguest explanation of brain problems and while apologetic that they could not do better, did not suggest that it mattered in any way or that they intended to pursue further knowledge. What participants did mention researching, as noted in earlier chapters, were diagnostic categories and information about the side effects of particular medications.

16. This pattern has been found in other qualitative research; see, for example, Malpass et al., "'Medication Career' or 'Moral Career'?," a "meta-ethnography" of studies of patient experiences of antidepressants.

17. Nor, of course, a medical treatment like an insulin shot to treat diabetes.

18. See, for example, DeSantis and Hane, "'Adderall Is Definitely Not a Drug.'" This is a theme in the DTC ads for medications to treat ADHD: "Schoolwork that matches his intelligence," as one ad for Adderall claimed, showing a boy holding up an exam with a B+ grade, while being energetically hugged by his joyous mother. Ad in author's personal collection.

19. There was the issue of diagnosis and mental disorder, but she dispatches these concerns in the manner of a "third condition."

20. See Metzl, *Prozac on the Couch,* chapter 5, for an interesting discussion of the "present tense" in Prozac narratives.

21. An observation of the sociologist Robert Bellah ("The Quest for the Self," 372) seems apposite here. He writes, "the ethos of American individualism seems deter-

mined more than ever to press ahead with the task of letting go of all criteria other than a radically private validation."

22. Marx and Engels, "Manifesto of the Communist Party," 476.

23. Bauman, *Liquid Modernity*, 7.

24. Ehrenberg, *The Weariness of the Self*, 185, original emphasis, 117. According to the sociologist Ulrich Beck (*Risk Society*, 135, 136), self-definition has been progressively transformed from something largely "given" to something largely "produced." Ordering one's biography is increasingly a project "placed in his or her own hands," which must be viewed and treated as "dependent on decisions" about virtually every aspect of one's life. It is a project for which each is responsible and held accountable.

25. Veroff, Douvan, and Kulka, *The Inner American*, 529.

26. Sennett, *The Corrosion of Character*, 9.

27. Bauman, *Liquid Modernity*, 126.

28. Bauman, *Liquid Times*, 4.

29. Rosa, *Social Acceleration*, 117. He also notes that the problem of falling behind is aggravated by constant change in the "decision landscape," which makes it increasingly difficult to make choices as it becomes ever harder to predict which of the potential options will turn out to be relevant and beneficial in the future.

30. Quoted in Schwartz, "Tyranny of Choice," 48.

31. Beck, *Risk Society*, 135.

32. Giddens, *Modernity and Self-Identity*.

33. On missing out, see Rosa, *Social Acceleration*, 184. He observes that because the same changes that permit accelerated realization of worldly possibilities also multiply the number of realizable options, the "degree of missing out" keeps expanding—"often in an exponential way." Reflecting this very dynamic, FOMO, or "fear of missing out," has become a widely used acronym.

34. Beck, *Risk Society*, 136.

35. De Zengotita, *Mediated*.

36. Riesman, *The Lonely Crowd*. What Riesman described are "ideal-types," analytical constructs assembled from certain general features and tendencies but which do not characterize anyone in particular. While he suggested that other-direction was becoming predominate, he also stressed that the other types (tradition- and inner-direction) persist in diminished frequency. The notion of ideal-type comes from Max Weber ("'Objectivity' in Social Science," 90, original emphasis). According to his definition, "An ideal type is formed by the one-sided *accentuation* of one or more points of view and by the synthesis of a great many diffuse, discrete, more or less present and occasionally absent *concrete individual* phenomena, which are arranged according to those one-sidedly emphasized viewpoints into a unified *analytical* construct (*Gedankenbild*). In its conceptual purity, this mental construct (*Gedankenbild*) cannot be found empirically anywhere in reality."

37. In a celebrated 1984 essay, the literary theorist Fredric Jameson ("Postmodernism," 62) argued that theoretical discourses in art, architecture, music, and literary criticism had repudiated "fundamental depth models" of subjectivity, including the "hermeneutic model of the inside and the outside" and the "existential model of authenticity and inauthenticity." In these and other models, "depth is replaced by surface, or by multiple surfaces." In the same year, the historian Christopher Lasch (*The Minimal Self*, 154, 153) described the "minimalist aesthetic" of art and literature that arose in the 1950s and 1960s as a "flight from selfhood," a "turning away

from the interior world," and a renouncing of what a novelist Lasch quotes calls the "old myths of 'depth.'"

38. Lasch, *The Minimal Self*, 258.

39. In a diary study of adolescent American girls, the historian Joan Jacobs Brumberg (*The Body Project*, xxi) gives this illustrative example from 1892: "'Resolved, not to talk about myself or feelings. To think before speaking. To work seriously. To be self restrained in conversation and actions. Not to let my thoughts wander. To be dignified. Interest myself more in others.'"

40. Bauman *Liquid Times*, 3.

41. According to one interview participant, Lauren, "I think people like to avoid thinking about things that bother them, or they like to avoid things that cause them stress. So I think it's easier to take medication, because then you don't have to go into all of those issues that cause pain." In a large-scale study of attitudes toward depression and depression treatment, researchers asked if "counseling brings up too many bad feelings such as anger or sadness." Nearly 43 percent agreed or agreed strongly with the statement. Givens et al., "Ethnicity and Preferences for Depression Treatment," 186.

42. Lasch, *The Minimal Self*, 15.

43. Thomson, *In Conflict No Longer*, 107.

44. Conley, *Elsewhere, U.S.A.*

45. Ehrenberg, *The Weariness of the Self*, 116.

46. Of course, there are social movements of various kinds that criticize the social order. But their starting point is normally just this picture of society as a space of reflexive self-designing. Their aim is not to critique that picture but to promote it, to make it more inclusive of the marginalized. See studies in Davis, "Narrative and Social Movements."

47. Marx held a similar view. See the discussion in Nieman, *Evil in Modern Thought*, 104–6. So did others, following Weber, in the interpretive tradition. See, for example, Geertz, *Interpretation of Cultures*, and Berger, *The Sacred Canopy*.

48. All quotes in this paragraph are from Weber, "The Social Psychology of the World Religions," 275. See also the chapter "Theodicy, Salvation, and Rebirth," in Weber, *The Sociology of Religion*, 138–50.

49. Weber, "The Social Psychology of the World Religions," 276.

50. Murdoch, *Sovereignty of Good*, 55.

51. Wilkinson, *Suffering*, 29. See also Berger, *The Sacred Canopy*.

52. Comaroff, "Medicine: Symbol and Ideology" 55.

53. Gordon, "Tenacious Assumptions in Western Medicine," 37, 40. See Murdoch, *Sovereignty of Good* (chapter 1), for much the same picture of the human that dominates modern moral philosophy.

CONCLUSION

1. Keller, "Whole Bodies," 357.

2. Flanagan, *The Problem of the Soul*, xii.

3. Crick, *The Astonishing Hypothesis*, 3.

4. Churchland, *Touching a Nerve*, 32.

5. Rosenberg, "Why You Don't Know Your Own Mind."

6. Kihlstrom, "The Automaticity Juggernaut." According to psychologist Timothy Wilson (*Strangers to Ourselves*, 6), the role of consciousness in human experience is something like the "size of a snowball" on top of an iceberg.

7. Wilson, *Strangers to Ourselves*.

8. Hall, "Introduction: Who Needs 'Identity'?," 6.

9. In an article titled "Inwardness," for *The Point* magazine, English professor Lisa Ruddick observes, "According to academic Marxist theory, inwardness itself is quietistic; . . . it is really the larger social forces that determine who we are and what the nature is of our subjectivity, and along with this comes a depreciation of everything modern and Western, including supposedly, the individuated self or the introspective self." https://thepointmag.com/2014/criticism/inwardness.

10. See, for instance, Churchland, *Touching a Nerve*, and Wilson, *The Meaning of Human Existence*.

11. Nagel, *Mortal Questions*, 38.

12. Murdoch, *Sovereignty of Good*, 16.

13. As philosopher Alva Noë observes in *Out of Our Heads* (6), "what needs to be kept clearly in focus is that the neuroscientists . . . have really only succeeded in replacing one mystery with another."

14. See the discussion in Metzl (*Prozac on the Couch*, 176) of authors who use the neurobiological as a critique of the confining nature of psychotherapy and, especially, psychoanalysis. He shows that it was combined with the hope that antidepressant medication would help them become "productively hyperthymic, optimistic, decisive, and quick of thought in a culture that marks those who are not as pathological." The neurobiological, in other words, draws them more deeply into the contemporary regime of the self. For a review of some of the postmodern arguments for the "generative" possibilities of brain talk for progressive politics, see Pitts-Taylor, "The Plastic Brain."

15. Rose, *Politics of Life Itself*, 26.

16. The quote is from Alexander Mitscherlich, cited by the philosopher Jürgen Habermas, *The Future of Human Nature*, 5.

17. Murdoch, *Sovereignty of Good*, 22.

18. These are the emotions that Taylor says have "subject-referring" properties. They "incorporate a sense of what is important to us in our lives as subjects" and so involve our self-worth and vision of the good life. Taylor, *Human Agency and Language*, 54, 60.

19. Archer, *Being Human*, 193. I take this definition to be very close to or identical with philosopher Robert C. Roberts's definition of emotions as "concern-based construals," by which he means "'takes' on or ways of 'seeing' situations (in the 'world,' not in the body), some crucial element(s) of which the subject cares about" (Roberts, "Emotions and Culture," 22). Similarly, the philosopher Talbot Brewer ("On Alienated Emotions," 285–86) argues that "emotions are vivid 'seemings' of the evaluative contours of one's changing circumstances, and that they serve to set the stage for evaluative understanding and action. . . . Emotions . . . are about the world, and they present us with a picture of the value or importance of various features of our circumstances."

20. Murdoch, *Sovereignty of Good*, 22.

21. According to psychoanalysts, a threatening emotion may set off an alarm in us, what they call "signal anxiety," leading to its unconscious repression.

22. Just such a transformed understanding is the central argument of Ivan Illich's *Medical Nemesis*.

# BIBLIOGRAPHY

Ahn, Woo-kyoung, Caroline C. Proctor, and Elizabeth H. Flanagan. "Mental Health Clinicians' Beliefs about the Biological, Psychological, and Environmental Bases of Mental Disorders." *Cognitive Science* 33 (2009): 147–82.

Aikin, Kathryn J., John L. Swasy, and Amie C. Braman. "Patient and Physician Attitudes and Behaviors Associated with DTC Promotion of Prescription Drugs—Summary of FDA Survey Research Results." Washington, DC: US Department of Health and Human Services, November 19, 2004.

American Psychiatric Association. *Diagnostic and Statistical Manual of Mental Disorders.* 1st ed. (DSM-I). Washington, DC: American Psychiatric Association, 1952.

American Psychiatric Association. *Diagnostic and Statistical Manual of Mental Disorders.* 2nd ed. (DSM-II). Washington, DC: American Psychiatric Association, 1968.

American Psychiatric Association. *Diagnostic and Statistical Manual of Mental Disorders.* 3rd ed. (DSM-III). Washington, DC: American Psychiatric Association, 1980.

American Psychiatric Association. *Diagnostic and Statistical Manual of Mental Disorders,* 4th ed., text revision (DSM-IV-TR). Washington, DC: American Psychiatric Association, 2000.

American Psychiatric Association. *Diagnostic and Statistical Manual of Mental Disorders.* 5th ed. (DSM-5). Washington, DC: American Psychiatric Publishing, 2013.

Archer, Margaret S. *Being Human: The Problem of Agency.* Cambridge: Cambridge University Press, 2000.

Armstrong, David. "The Patient's View." *Social Science and Medicine* 18 (1984): 737–44.

Austin, J. L. *How to Do Things with Words.* Cambridge, MA: Harvard University Press, 1962.

Bakhtin, M. M. *Speech Genres and Other Late Essays.* Austin: University of Texas Press, 1986.

Barber, Charles. *Comfortably Numb: How Psychiatry Is Medicating a Nation.* New York: Vintage Books, 2009.

Baum, Carlene, Dianne L. Kennedy, Deanne E. Knapp, John P. Juergens, and Gerald A. Faich. "Prescription Drug Use in 1984 and Changes over Time." *Medical Care* 26 (1988): 105–14.

Bauman, Zygmunt. *Liquid Modernity.* Malden, MA: Blackwell, 2000.

Bauman, Zygmunt. *Liquid Times: Living in an Age of Uncertainty.* Cambridge: Polity Press, 2007.

Beck, Ulrich. *Risk Society: Towards a New Modernity.* London: Sage, 1992.

Becker, Howard S. "Becoming a Marihuana User." *American Journal of Sociology* 59 (1953): 235–42.

Becker, Howard S. "History, Culture and Subjective Experience: An Exploration of the Social Bases of Drug-Induced Experiences." *Journal of Health and Social Behavior* 8 (1967): 163–76.

Bell, Daniel. *The Cultural Contradictions of Capitalism.* New York: Basic Books, 1976.

Bellah, Robert N. "The Quest for the Self: Individualism, Morality, Politics." In *Interpretive Social Science: A Second Look,* edited by Paul Rabinow and William M. Sullivan, 365–83. Berkeley: University of California Press, 1987.

Berger, Peter L. *The Sacred Canopy: Elements of a Sociological Theory of Religion.* Garden City, NY: Anchor Books, 1969.

Berger, Peter L., and Thomas Luckmann. *The Social Construction of Reality.* New York: Anchor Books, 1966.

Berlin, Isaiah. *Four Essays on Liberty.* Oxford: Oxford University Press, 1969.

Bilderbeck, A. C., K. E. Saunders, J. Price, and G. M. Goodwin. "Psychiatric Assessment of Mood Instability: Qualitative Study of Patient Experience." *British Journal of Psychiatry* 204 (2014): 234–39.

Blackwell, Barry. "Minor Tranquilizers: Use, Misuse or Overuse?" *Psychosomatics* 16 (1975): 28–31.

Blackwell, Barry. "Psychotropic Drugs in Use Today: The Role of Diazepam in Medical Practice." *Journal of the American Medical Association* 225 (1973): 1637–41.

Boltanski, Luc. *Distant Suffering: Morality, Media and Politics.* Translated by Graham D. Burchell. Cambridge: Cambridge University Press, 1999.

Boltanski, Luc. *The Foetal Condition: A Sociology of Engendering and Abortion.* Translated by Catherine Porter. Cambridge: Polity, 2013.

Boltanski, Luc. *Love and Justice as Competences.* Translated by Catherine Porter. Cambridge: Polity, 2012.

Bourne, Edmund J. *The Anxiety and Phobia Workbook.* 3rd ed. Oakland, CA: New Harbinger Publications, 2000.

Braslow, Joel T. *Mental Ills and Bodily Cures: Psychiatric Treatment in the First Half of the Twentieth Century.* Berkeley: University of California Press, 1997.

Brewer, Talbot. "On Alienated Emotions." In *Morality and the Emotions,* edited by Carla Bagnoli, 275–98. New York: Oxford University Press, 2011.

Brownlee, Shannon. "Mysteries of the Mind." *Washington Post,* October 3, 2004, B01.

Bruckner, Pascal. *Perpetual Euphoria: On the Duty to Be Happy.* Princeton, NJ: Princeton University Press, 2010.

Brumberg, Joan Jacobs. *The Body Project: An Intimate History of American Girls.* New York: Vintage Books, 1997.

Bruner, Jerome. *Acts of Meaning.* Cambridge, MA: Harvard University Press, 1990.

Bury, Michael. *Health and Illness in a Changing Society.* New York: Routledge, 1997.

Canguilhem, Georges. *On the Normal and the Pathological.* Dordrecht: D. Reidel, 1978.

Carlat, Daniel. *Unhinged: The Trouble with Psychiatry—a Doctor's Revelations about a Profession in Crisis.* New York: Free Press, 2010.

Cassell, Eric J. *The Nature of Suffering and the Goals of Medicine.* New York: Oxford University Press, 1991.

Cassell, Eric J. "Recognizing Suffering." *Hastings Center Report* 21, no. 3 (1991): 24–31.

Castoriadis, Cornelius. *The Imaginary Institution of Society.* Cambridge: Polity, 1987.

Chesler, Phyllis. *Women and Madness.* Garden City, NY: Doubleday, 1972.

Churchland, Patricia. *Touching a Nerve: The Self as Brain.* New York: W. W. Norton, 2013.

Clarke, Adele E. "From the Rise of Medicine to Biomedicalization: U.S. Healthscapes and Iconography, circa 1890–Present." In *Biomedicalization: Technoscience, Health, and Illness in the U.S.*, edited by Adele E. Clarke, Laura Mamo, Jennifer Ruth Fosket, Jennifer R. Fishman, and Janet K. Shim, 104–46. Durham, NC: Duke University Press, 2010.

Clarke, Juanne, and Adele Gawley. "The Triumph of Pharmaceuticals: The Portrayal of Depression from 1980 to 2005." *Administration and Policy in Mental Health* 36 (2009): 91–101.

Cohen, Eric P. "Direct-to-the-Public Advertisement of Prescription Drugs." *New England Journal of Medicine* 318 (1988): 373–76.

Collins, Harry. *Tacit and Explicit Knowledge*. Chicago: University of Chicago Press, 2010.

Comaroff, Jean. "Medicine: Symbol and Ideology." In *The Problem of Medical Knowledge: Examining the Social Construction of Medicine*, edited by Peter Wright and Andrew Treacher, 49–68. Edinburgh: Edinburgh University Press, 1982.

Conley, Dalton. *Elsewhere, U.S.A.* New York: Vintage Books, 2009.

Connolly, William E. "The Human Predicament." *Social Research* 76 (2009): 1121–40.

Conrad, Peter. *The Medicalization of Society: On the Transformation of Human Conditions into Treatable Disorders*. Baltimore, MD: Johns Hopkins University Press, 2007.

Conrad, Peter, and Deborah Potter. "From Hyperactive Children to ADHD Adults: Observations on the Expansion of Medical Categories." *Social Problems* 47 (2000): 559–82.

Cooper, Rachel. *Classifying Madness: A Philosophical Examination of the Diagnostic and Statistical Manual of Mental Disorders*. Dordrecht: Springer, 2005.

Cooperstock, Ruth, and Henry L. Lennard. "Some Social Meanings of Tranquilizer Use." *Sociology of Health and Illness* 1 (1979): 331–47.

Coppen, Alec. "The Biochemistry of Affective Disorders." *British Journal of Psychiatry* 113 (1967): 1237–64.

Cottle, Michelle. "Selling Shyness." *New Republic*, August 2, 1999, 24–29.

Coulter, Jeff. "Beliefs and Practical Understanding." In *Everyday Language: Studies in Ethnomethodology*, edited by George Psathas, 163–86. New York: Irvington Publishing, 1979.

Crick, Francis. *The Astonishing Hypothesis: The Scientific Search for the Soul*. New York: Charles Scribner's Sons, 1994.

Critser, Greg. *Generation Rx*. Boston: Houghton Mifflin, 2005.

Crossley, Nick. *Contesting Psychiatry: Social Movements in Mental Health*. New York: Routledge, 2006.

Crossley, Nick. "Prozac Nation and the Biochemical Self: A Critique." In *Debating Biology: Sociological Reflections on Health, Medicine and Society*, edited by Simon J. Williams, Lynda Birke, and Gillian A. Bendelow, 245–58. London: Routledge, 2003.

Cruikshank, Barbara. *The Will to Empower: Democratic Citizens and Other Subjects*. Ithaca, NY: Cornell University Press, 1999.

Cumming, Elaine, and John Cumming. *Closed Ranks: An Experiment in Mental Health Education*. Cambridge, MA: Harvard University Press, 1957.

Cumming, John, and Elaine Cumming. "Mental Health Education in a Canadian Community." *Sociological Practice* 8 (1990): 96–115. First published 1955.

D'Arcy, Carl, and Joan Brockman. "Changing Public Recognition of Psychiatric Symptoms? Blackfoot Revisited." *Journal of Health and Social Behavior* 17 (1976): 302–10.

Dave, Dhaval M. "Effects of Pharmaceutical Promotion: A Review and Assessment." Working Paper 18830. National Bureau of Economic Research, February 2013.

Davis, Joseph E. *Accounts of Innocence: Sexual Abuse, Trauma, and the Self*. Chicago: University of Chicago Press, 2005.

Davis, Joseph E. "Adolescents and the Pathologies of the Achieving Self." *Hedgehog Review* 11, no. 1 (Spring 2009): 37–49.

Davis, Joseph E. "The Commodification of Self." *Hedgehog Review* 5, no. 2 (Summer 2003): 41–49.

Davis, Joseph E. "Emotions as Commentaries on Cultural Norms." In *The Emotions and Cultural Analysis*, edited by Ana Marta González, 31–49. Burlington, VT: Ashgate, 2012.

Davis, Joseph E. "Medicalization, Social Control, and the Relief of Suffering." In *The New Blackwell Companion to Medical Sociology*, edited by William C. Cockerham, 211–41. Malden, MA: Blackwell Publishers, 2010.

Davis, Joseph E. "Narrative and Social Movements: The Power of Stories." In *Stories of Change: Narrative and Social Movements*, edited by in Joseph E. Davis, 3–29. Albany: State University of New York Press, 2002.

Davis, Joseph E. "Post-Prozac Pathography." *Hedgehog Review* 16, no. 3 (Fall 2014): 10–12.

Davis, Joseph E. "Reductionist Medicine and Its Cultural Authority." In *To Fix or to Heal: Patient Care, Public Health, and the Limits of Biomedicine*, edited by Joseph E. Davis and Ana Marta González, 33–62. New York: New York University Press, 2016.

Davis, Joseph E. "Suffering, Pharmaceutical Advertising, and the Face of Mental Illness." *Hedgehog Review* 8, no. 3 (Fall 2006): 62–77.

Deferio, Joseph J., Tomer T. Levin, Judith Cukor, Samprit Banerjee, Rozan Abdulrahman, Amit Sheth, Neel Mehta, and Jyotishman Pathak. "Using Electronic Health Records to Characterize Prescription Patterns: Focus on Antidepressants in Nonpsychiatric Outpatient Settings." *JAMIA Open* 1 (2018): 233–45.

DeGloma, Thomas. *Seeing the Light: The Social Logic of Personal Discovery*. Chicago: University of Chicago Press, 2014.

DeGrandpre, Richard. *The Cult of Pharmacology: How America Became the World's Most Troubled Drug Culture*. Durham, NC: Duke University Press, 2006.

DeSantis, Alan, and Audrey Curtis Hane. "'Adderall Is Definitely Not a Drug': Justifications for the Illegal Use of ADHD Stimulants." *Substance Use & Misuse* 45 (2010): 31–46.

De Zengotita, Thomas. *Mediated: How the Media Shapes Your World and the Way You Live in It*. New York: Bloomsbury, 2005.

Douglas, Mary. *Purity and Danger: An Analysis of Concepts of Pollution and Taboo*. Baltimore: Pelican Books, 1970.

Douglas, Virginia I. "Stop, Look and Listen: The Problem of Sustained Attention and Impulse Control in Hyperactive and Normal Children." *Canadian Journal of Behavioural Science* 4 (1972): 259–82.

du Gay, Paul. "Against 'Enterprise' (but Not Against 'Enterprise,' for That Would Make No Sense)." *Organization* 11, no. 1 (2004): 37–57.

Dumit, Joseph. *Drugs for Life: How Pharmaceutical Companies Define Our Health*. Durham, NC: Duke University Press, 2012.

Ehrenberg, Alain. *The Weariness of the Self: Diagnosing the History of Depression in the Contemporary Age*. Montreal and Kingston: McGill-Queen's University Press, 2010. First published in French 1998.

Elias, Norbert. *The Civilizing Process*. New York: Urizen Books, 1978.

Engel, George L. "The Need for a New Medical Model: A Challenge for Biomedicine." *Science* 196 (April 8, 1977): 129–36.

Ewen, Stuart. *All Consuming Images: The Politics of Style in Contemporary Culture*. New York: Basic Books, 1988.

Eysenck, H. J. *The Biological Basis of Personality*. Springfield, IL: Charles C. Thomas, 1967.

Feighner, John P. "The Advent of the Feighner Criteria." *Current Contents* 43 (October 23, 1989): 14.

Feighner, John P., Eli Robins, Samuel B. Guze, Robert A. Woodruff Jr., George Winokur, and Rodrigo Munoz. "Diagnostic Criteria for Use in Psychiatric Research." *Archives of General Psychiatry* 26 (1972): 57–63.

Fieve, Ronald R. *Moodswing: The Third Revolution in Psychiatry*. New York: William Morrow, 1975.

Finniss, Damien G., Ted Kaptchuk, Franklin Miller, and Fabrizio Benedetti. "Biological, Clinical, and Ethical Advances of Placebo Effects." *Lancet* 375 (2010): 686–95.

Flanagan, Owen. *The Problem of the Soul: Two Visions of Mind and How to Reconcile Them*. New York: Basic Books, 2002.

Foucault, Michel. "On the Genealogy of Ethics: An Overview of Work in Progress." In *The Essential Foucault: Selections from the Essential Works of Foucault, 1954–1984*, edited by Paul Rabinow and Nikolas Rose, 102–25. New York: The New Press, 2003.

Frances, Allen. *Saving Normal: An Insider's Revolt against Out-of-Control Psychiatric Diagnosis, DSM-5, Big Pharma, and the Medicalization of Ordinary Life*. New York: HarperCollins, 2013.

Frances, Allen, Avram H. Mack, Ruth Ross, and Michael B. First. "The DSM-IV Classification and Psychopharmacology." In *Psychopharmacology: The Fourth Generation of Progress*, edited by Floyd E. Bloom and David J. Kupfer, 823–28. New York: Raven Press, 1995.

Frank, Jerome D., and Julia B. Frank. *Persuasion and Healing: A Comparative Study of Psychotherapy*. 3d ed. Baltimore, MD: Johns Hopkins University Press, 1991.

Friedan, Betty. *The Feminine Mystique*. New York: Norton, 1963.

Frye, Mark A., Terence A. Ketter, Gabriele S. Leverich, Teresa Huggins, Caprice Lantz, Kirk D. Denicoff, and Robert M. Post. "The Increasing Use of Polypharmacotherapy for Refractory Mood Disorders: 22 Years of Study." *Journal of Clinical Psychiatry* 61 (2000): 9–15.

Furukawa, Toshi A., Andrea Cipriani, Lauren Z. Atkinson, Stefan Leucht, Yusuke Ogawa, Nozomi Takeshima, Yu Hayasaka, Anna Chaimani, and Georgia Salanti. "Placebo Response Rates in Antidepressant Trials: A Systematic Review of Published and Unpublished Double-Blind Randomised Controlled Studies." *Lancet Psychiatry* 3 (2016): 1059–66.

Gardos, George, Andras Perenyi, and Jonathan O. Cole. "Polypharmacy Revisited." *McLean Hospital Journal* 5 (1980): 178–95.

Garfield, Sol L. "Research on Client Variables in Psychotherapy." In *Handbook of Psychotherapy and Behavior Change*, 3d ed., edited by Sol L. Garfield and Allen E. Bergin, 213–56. New York: John Wiley & Sons, 1986.

Garnier-Dykstra, Laura M., Kimberly M. Caldeira, Kathryn B. Vincent, Kevin E. O'Grady, and Amelia M. Arria. "Nonmedical Use of Prescription Stimulants during College: Four-Year Trends in Exposure Opportunity, Use, Motives, and Sources." *Journal of American College Health* 60 (2012): 226–34.

Geertz, Clifford. *The Interpretation of Cultures*. New York: Basic Book, 1973.

Giddens, Anthony. *Central Problems in Social Theory: Action, Structure and Contradiction in Social Analysis*. London: Macmillan Press, 1979.

Giddens, Anthony. "Living in a Post-Traditional Society." In *Reflexive Modernization: Politics, Tradition and Aesthetics in the Modern Social Order*, edited by Ulrich Beck, Anthony Giddens and Scott Lash, 56–109. Cambridge: Polity Press, 1994.

Giddens, Anthony. *Modernity and Self-Identity: Self and Society in the Late Modern Age*. Stanford, CA: Stanford University Press, 1991.

Givens, Jane L., Thomas K. Houston, Benjamin W. Van Voorhees, Daniel E. Ford, and Lisa A. Cooper. "Ethnicity and Preferences for Depression Treatment." *General Hospital Psychiatry* 29 (2007): 182–91.

Goffman, Erving. *Interaction Ritual: Essays on Face-to-Face Behavior*. New York: Pantheon, 1967.

Goffman, Erving. *The Presentation of Self in Everyday Life*. New York: Doubleday, 1959.

Goffman, Erving. *Stigma: Notes on the Management of Spoiled Identity*. New York: Simon and Schuster, 1963.

Goldhamer, Herbert, and Andrew W. Marshall. *Psychosis and Civilization: Two Studies in the Frequency of Mental Disease*. Glencoe, IL: Free Press, 1953.

Goldman, Howard H. "Progress in the Elimination of the Stigma of Mental Illness." *American Journal of Psychiatry* 167 (2010): 1289–90.

Goldstein, Benjamin, and Francine Rosselli. "Etiological Paradigms of Depression: The Relationship between Perceived Causes, Empowerment, Treatment Preferences, and Stigma." *Journal of Mental Health* 12 (2003): 551–63.

Goleman, Daniel. *Emotional Intelligence*. New York: Bantam Books, 1995.

Gordon, Deborah R. "Tenacious Assumptions in Western Medicine." In *Biomedicine Examined*, edited by Margaret Lock and Deborah R. Gordon, 19–56. Dordrecht: Kluwer Academic Publishers, 1988.

Gottlieb, Lori. "The Branding Cure." *New York Times Magazine*, November 25, 2012, MM36.

Gray, Alison J. "Stigma in Psychiatry." *Journal of the Royal Society of Medicine* 95 (2002): 72–76.

Gray, J. A. "Anxiety and the Brain: Not by Neurochemistry Alone." *Psychological Medicine* 9 (1979): 605–9.

Grebb, Jack A., and Arvid Carlsson. "Introduction and Considerations for a Brain-Based Diagnostic System in Psychiatry." In *Kaplan and Sadock's Comprehensive Textbook of Psychiatry*, 9th ed., edited by Benjamin J. Sadock, Virginia A. Sadock, and Pedro Ruiz, 1: 1–5. Philadelphia: Wolters Kluwer Health/Lippincott Williams and Wilkins, 2009.

Greco, Monica. "Psychosomatic Subjects and the 'Duty to Be Well': Personal Agency within Medical Rationality." *Economy and Society* 22 (1993): 357–72.

Greco, Monica. "Thinking beyond Polemics: Approaching the Health Society through Foucault." *Österreichische Zeitschrift für Soziologie* 34, no. 2 (2009): 13–27.

Greene, Jeremy A. *Prescribing by Numbers: Drugs and the Definition of Disease*. Baltimore, MD: Johns Hopkins University Press, 2007.

Gross, Martin L. *The Psychological Society: A Critical Analysis of Psychiatry, Psychotherapy, Psychoanalysis and the Psychological Revolution*. New York: Random House, 1978.

Grow, Jean M., Jin Seong Park, and Xiaoqi Han. "'Your Life Is Waiting!' Symbolic Meanings in Direct-to-Consumer Antidepressant Advertising." *Journal of Communication Inquiry* 30 (2006): 163–88.

Guignon, Charles. *On Being Authentic*. New York: Routledge, 2004.

Gurin, Gerald, Joseph Veroff, and Sheila Feld. *Americans View Their Mental Health*. New York: Basic Books, 1960.

Habermas, Jürgen. *The Future of Human Nature*. Cambridge: Polity, 2003.

Hacking, Ian. *The Taming of Chance*. Cambridge: Cambridge University Press, 1990.

Hale, Nathan G., Jr. *The Rise and Crisis of Psychoanalysis in the United States: Freud and the Americans, 1917–1985*. New York: Oxford University Press, 1995.

Hall, Stuart. "Introduction: Who Needs 'Identity'?" In *Questions of Cultural Identity*, edited by Stuart Hall and Paul DuGay, 1–17. Thousand Oaks, CA: Sage, 1996.

Haslam, Nick. "Folk Psychiatry: Lay Thinking about Mental Disorder." *Social Research* 70, no. 2 (2003): 621–44.

Hawkins, Anne Hunsaker. *Reconstructing Illness: Studies in Pathography*, 2nd ed. West Lafayette, IN: Purdue University Press, 1999.

Healy, David. *The Antidepressant Era*. Cambridge, MA: Harvard University Press, 1997.

Healy, David. *Let Them Eat Prozac: The Unhealthy Relationship between the Pharmaceutical Industry and Depression*. New York: New York University Press, 2004.

Heatherington, Laurie, Stanley B. Messer, Lynne Angus, Timothy J. Strauman, Myrna L. Friedlander, and Gregory G. Kolden. "The Narrowing of Theoretical Orientations in Clinical Psychology Doctoral Training." *Clinical Psychology: Science and Practice* 19 (2012): 362–74.

Henig, Robin Marantz. "Valium's Contribution to Our New Normal." *New York Times*, September 29, 2012.

Herzberg, David. *Happy Pills in America: From Miltown to Prozac*. Baltimore, MD: Johns Hopkins University Press, 2009.

Hochschild, Arlie Russell. "Emotion Work, Feeling Rules, and Social Structure." *American Journal of Sociology* 85 (1979): 551–75.

Hoffman, Diane M. "Raising the Awesome Child." *Hedgehog Review* 15, no. 3 (Fall 2013): 30–41.

Hoffman, Jerome R., and Michael Wilkes. "Direct to Consumer Advertising of Prescription Drugs: An Idea Whose Time Should Not Come." *British Medical Journal* 318 (1999): 1301–2.

Hollingshead, August B., and Fredrick C. Redlich. *Social Class and Mental Illness: A Community Study*. New York: Wiley, 1958.

Hollister, Leo E. "Drugs for Emotional Disorders." *Journal of the American Medical Association* 234 (1975): 942–47.

Hollister, Leo E. "Psychopharmacology in Its Second Generation." *Military Medicine* 141 (1976): 371–75.

Horwitz, Allan V. *The Social Control of Mental Illness*. New York: Academic Press, 1982.

Horwitz, Allan V. "Transforming Normality into Pathology: The DSM and the Outcomes of Stressful Social Arrangements." *Journal of Health and Social Behavior* 48 (2007): 211–22.

Horwitz, Allan V., and Jerome C. Wakefield. *The Loss of Sadness: How Psychiatry Transformed Normal Sorrow into Depressive Disorder*. New York: Oxford University Press, 2007.

Howie, LaJeana D., Patricia N. Pastor, and Susan L. Lukacs. "Use of Medication Prescribed for Emotional or Behavioral Difficulties among Children Aged 6–17 Years in the United States, 2011–2012." *NCHS Data Brief*, no. 148. Hyattsville, MD: National Center for Health Statistics, 2014.

Hyman, Steven E. "The Daunting Polygenicity of Mental Illness: Making a New Map." *Philosophical Transactions of the Royal Society B Biological Sciences* 373 (2018): 20170031.

Hyman, Steven E. "The Diagnosis of Mental Disorders: The Problem of Reification." *Annual Review of Clinical Psychology* 6 (2010): 155–79.

Illich, Ivan. *Medical Nemesis: The Expropriation of Health*. New York: Random House, 1976.

Illouz, Eva. *Saving the Modern Soul: Therapy, Emotions, and the Culture of Self-Help.* Berkeley: University of California Press, 2008.

Insel, Thomas R., and Remi Quirion. "Psychiatry as a Clinical Neuroscience Discipline." *Journal of the American Medical Association* 294 (November 2, 2005): 2221–4.

Iselin, Marie-Geneviève, and Michael E. Addis. "Effects of Etiology on Perceived Helpfulness of Treatments for Depression." *Cognitive Therapy and Research* 27 (2003): 205–22.

Itkowitz, Colby. "Unwell and Unashamed." *Washington Post,* June 2, 2016, A1, A12–A13.

Jameson, Fredric. "Postmodernism, or The Cultural Logic of Late Capitalism." *New Left Review* 146 (July–August 1984): 53–92.

Jasanoff, Shelia. "Future Imperfect: Science, Technology, and the Imaginations of Modernity." In *Dreamscapes of Modernity: Sociotechnical Imaginaries and the Fabrication of Power,* edited by Shelia Jasanoff and Sang-Hyun Kim, 1–33. Chicago: University of Chicago Press, 2015.

Johnson, Spencer. *Who Moved My Cheese? An Amazing Way to Deal with Change in Your Work and in Your Life.* New York: G. P. Putnam's Sons, 1998.

Joint Commission on Mental Illness and Health. *Action for Mental Health.* New York: Basic Books, 1961.

Jones, Nelson F., Marvin W. Kahn, and John M. Macdonald. "Psychiatric Patients' Views of Mental Illness, Hospitalization and Treatment." *Journal of Nervous and Mental Disease* 136 (1963): 82–87.

Kadison, Richard, and Theresa Foy DiGeronimo. *College of the Overwhelmed: The Campus Mental Health Crisis and What to Do about It.* San Francisco: Jossey-Bass, 2004.

Kadushin, Charles. *Why People Go to Psychiatrists.* New York: Atherton Press, 1969.

Kandel, Eric. "A New Intellectual Framework for Psychiatry." *American Journal of Psychiatry* 155 (1998): 457–69.

Kantor, Elizabeth D., Colin D. Rehm, Jennifer S. Haas, Andrew T. Chan, and Edward L. Giovannucci. "Trends in Prescription Drug Use among Adults in the United States from 1999–2012." *Journal of the American Medical Association* 314 (2015): 1818–31.

Karp, David A. *Is It Me or My Meds? Living with Antidepressants.* Cambridge, MA: Harvard University Press, 2006.

Karp, David A. *Speaking of Sadness: Depression, Disconnection, and the Meaning of Illness.* New York: Oxford University Press, 1996.

Kaufman, Joanne. "Think You're Seeing More Drug Ads on TV? Well, You Are." *New York Times,* December 25, 2017, B4.

Keller, Evelyn Fox. "Whole Bodies, Whole Persons? Cultural Studies, Psychoanalysis, and Biology." In *Subjectivity: Ethnographic Investigations,* edited by João Biehl, Byron Good, and Arthur Kleinman, 352–61. Berkeley: University of California Press, 2007.

Kendler, Kenneth S. "Explanatory Models for Psychiatric Illness." *American Journal of Psychiatry* 165 (2008): 695–702.

Kessler, Ronald C., Olga Demler, Richard G. Frank, Mark Olfson, Harold Alan Pincus, Ellen E. Walters, Philip Wang, Kenneth B. Wells, and Alan M. Zaslavsky. "Prevalence and Treatment of Mental Disorders, 1990 to 2003." *New England Journal of Medicine* 352 (2005): 2515–23.

Kessler, Ronald C., Kathleen R. Merikangas, Patricia Berglund, William W. Eaton, Doreen S. Koretz, and Ellen E. Walters. "Mild Disorders Should Not Be Eliminated from the DSM-V." *Archives of General Psychiatry* 60 (2003): 1117–22.

Kierkegaard, Søren. *The Sickness unto Death.* Princeton, NJ: Princeton University Press, 1980. First published 1849.

Kihlstrom, John F. "The Automaticity Juggernaut—or, Are We Automatons after All?" In *Are We Free? Psychology and Free Will*, edited by John Baer, James C. Kaufman, and Roy F. Baumeister, 155–80. New York: Oxford University Press, 2008.

Kirmayer, Laurence J. "Mind and Body as Metaphors: Hidden Values in Biomedicine." In *Biomedicine Examined*, edited by Margaret Lock and Deborah R. Gordon, 57–93. Dordrecht: Kluwer Academic Publishers, 1988.

Klein, Donald F. "Delineation of Two Drug-Responsive Anxiety Syndromes." *Psychopharmacologia* 5 (1964) 397–408.

Klein, Donald F., Charlotte M. Zitrin, and Margaret Woerner. "Antidepressants, Anxiety, Panic, and Phobia." In *Psychopharmacology: A Generation of Progress*, edited by Morris A. Lipton, Alberto DiMascio, and Keith F. Killam, 1401–10. New York: Raven Press, 1978.

Kleinman, Arthur. *Rethinking Psychiatry: From Cultural Category to Personal Experience*. New York: Free Press, 1988.

Kline, Nathan S., ed. *Factors in Depression*. New York: Raven Press, 1974.

Kline, Nathan S. *From Sad to Glad: Kline on Depression*. New York: G. P. Putnam's Sons, 1974.

Kramer, Peter D. *Listening to Prozac*. New York: Viking Penguin, 1993.

Kramer, Peter D. *Ordinarily Well: The Case for Antidepressants*. New York: Farrar, Straus and Giroux, 2016.

LaBier, Douglas. "You've Gotta Think Like Google." *Washington Post*, November 11, 2008, F6.

Lader, M. "Benzodiazepines—the Opium of the Masses?" *Neuroscience* 3 (1978): 159–65.

Lane, Christopher. *Shyness: How Normal Behavior Became a Sickness*. New Haven, CT: Yale University Press, 2007.

Lasch, Christopher. *The Minimal Self*. New York: W. W. Norton, 1984.

Lebowitz, Matthew S. "Biological Conceptualizations of Mental Disorders among Affected Individuals: A Review of Correlates and Consequences." *Clinical Psychology: Science and Practice* 21 (2014): 67–83.

Lebowitz, Matthew S., John J. Pyun, and Woo-kyoung Ahn. "Biological Explanations of Generalized Anxiety Disorder: Effects on Beliefs about Prognosis and Responsibility." *Psychiatric Services* 65 (2014): 498–503.

Lennard, Henry L., Leon J. Epstein, Arnold Bernstein, and Donald C. Ransom. *Mystification and Drug Misuse: Hazards in Using Psychoactive Drugs*. San Francisco: Jossey-Bass, 1971.

Levenson, Hanna, and Donna Davidovitz. "Brief Therapy Prevalence and Training: A National Survey of Psychologists." *Psychotherapy* 37 (2000): 335–40.

Link, Bruce G., Jo C. Phelan, Michaeline Bresnahan, Ann Stueve, and Bernice A. Pescosolido. "Public Conceptions of Mental Illness: Labels, Causes, Dangerousness, and Social Distance." *American Journal of Public Health* 89 (1999): 1328–33.

Luhrmann, T. M. *Of Two Minds: The Growing Disorder in American Psychiatry*. New York: Alfred A. Knopf, 2000.

Luker, Kristin. *Salsa Dancing into the Social Sciences: Research in an Age of Info-glut*. Cambridge, MA: Harvard University Press, 2008.

Lupton, Deborah. *Medicine as Culture*. 2nd ed. London: Sage, 2003.

Mackey, Tim K., Raphael E. Cuomo, and Bryan A. Liang. "The Rise of Digital Direct-to-Consumer Advertising? Comparison of Direct-to-Consumer Advertising Expenditure Trends from Publicly Available Data Sources and Global Policy Implications." *BMC Health Services Research* 15 (2015): 236.

Malpass, Alice, Alison Shaw, Debbie Sharp, Fiona Walter, Gene Feder, Matthew Ridd, and David Kessler. "'Medication Career' or 'Moral Career'? The Two Sides of Managing Antidepressants: A Meta-Ethnography of Patients' Experience of Antidepressants." *Social Science and Medicine* 68 (2009): 154–68.

Manheimer, Dean I., Susan T. Davidson, Mitchell B. Balter, Glen D. Mellinger, Ira H. Cisin, and Hugh J. Parry. "Popular Attitudes and Beliefs about Tranquilizers." *American Journal of Psychiatry* 130 (1973): 1246–53.

March, Dana, and Gerald M. Oppenheimer. "Social Disorder and Diagnostic Order: The US Mental Hygiene Movement, the Midtown Manhattan Study and the Development of Psychiatric Epidemiology in the 20th Century." *International Journal of Epidemiology* 43, Suppl. 1 (2014): i29–i42.

Marcus, George E. "Introduction." In *Technoscientific Imaginaries: Conversations, Profiles, and Memoirs*, edited by George E. Marcus, 1–9. Chicago: University of Chicago Press, 1995.

Marshall, Kimball P., Zhanna Georgievskava, and Igor Georgievsky. "Social Reactions to Valium and Prozac: A Cultural Lag Perspective of Drug Diffusion and Adoption." *Research in Social and Administrative Pharmacy* 5 (2009): 94–107.

Marx, Karl, and Friedrich Engels. "Manifesto of the Communist Party." In *The Marx-Engels Reader*, 2nd ed., edited by Robert C. Tucker, 469–500. New York: Norton, 1978. First published 1848.

Mayes, Rick, and Allan V. Horwitz. "DSM-III and the Revolution in the Classification of Mental Illness." *Journal of the History of the Behavioral Sciences* 41 (2005): 249–67.

McCabe, Sean Esteban, Brady T. West, Christian J. Teter, and Carol J. Boyd. "Trends in Medical Use, Diversion, and Nonmedical Use of Prescription Medications among College Students from 2003 to 2013: Connecting the Dots." *Addictive Behaviors* 39 (2014): 1176–82.

McHugh, R. Kathryn, Sarah W. Whitton, Andrew D. Peckham, Jeffrey A. Welge, and Michael W. Otto. "Patient Preference for Psychological vs. Pharmacological Treatment of Psychiatric Disorders: A Meta-Analytic Review." *Journal of Clinical Psychiatry* 74 (2013): 595–602.

Medawar, Peter. "Victims of Psychiatry." *New York Review of Books*, January 23, 1975, 17.

Mehta, Sheila, and Amerigo Farina. "Is Being 'Sick' Really Better? Effect of the Disease View of Mental Disorder on Stigma." *Journal of Social and Clinical Psychology* 16 (1997): 405–19.

Mellinger, Glen D., Mitchell B. Balter, Dean I. Manheimer, Ira H. Cisin, and Hugh J. Parry. "Psychic Distress, Life Crisis, and Use of Psychotherapeutic Medications." *Archives of General Psychiatry* 35 (1978): 1045–52.

Mellinger, Glen D., Mitchell B. Balter, and Eberhard H. Uhlenhuth. "Prevalence and Correlates of the Long-Term Regular Use of Anxiolytics." *Journal of the American Medical Association* 251 (1984): 375–79.

Menninger, Karl, with Martin Mayman and Paul Pruyser. *The Vital Balance: The Life Process in Mental Health and Illness.* New York: Viking Press, 1963.

Metzl, Jonathan M. "'Mother's Little Helper': The Crisis of Psychoanalysis and the Miltown Resolution." *Gender & History* 15 (2003): 240–67.

Metzl, Jonathan Michel. *Prozac on the Couch: Prescribing Gender in the Era of Wonder Drugs.* Durham, NC: Duke University Press, 2003.

Metzl, Jonathan M., and Joni Angel, "Assessing the Impact of SSRI Antidepressants on Popular Notions of Women's Depressive Illness." *Social Science and Medicine* 58 (2004): 577–84.

Miller, Peter, and Nikolas Rose. "Mobilizing the Consumer: Assembling the Subject of Consumption." *Theory, Culture and Society* 14 (1997): 1–36.

Millett, Kate. *Sexual Politics*. Garden City, NY: Doubleday, 1970.

Mojtabai, Ramin, and Mark Olfson. "National Trends in Psychotherapy by Office-Based Psychiatrists." *Archives of General Psychiatry* 65 (2008): 962–70.

Mol, Annemarie. *The Logic of Care: Health and the Problem of Patient Choice*. New York: Routledge, 2008.

Moynihan, Ray, and Alan Cassels. *Selling Sickness*. New York: Nation Books, 2005.

Murdoch, Iris. *The Sovereignty of Good*. New York: Schocken Books, 1971.

Nagel, Thomas. *Mortal Questions*. New York: Cambridge University Press, 1979.

Nesse, Randolph M., and Dan J. Stein, "Towards a Genuinely Medical Model for Psychiatric Nosology." *BMC Medicine* 10 (2012): 5–13.

Nichter, Mark. "The Mission within the Madness: Self-Initiated Medicalization as Expression of Agency." In *Pragmatic Women and Body Politics*, edited by Margaret Lock and Patricia A. Kaufert, 327–53. Cambridge: Cambridge University Press, 1998.

Neiman, Susan. *Evil in Modern Thought: An Alternative History of Philosophy*. Princeton, NJ: Princeton University Press, 2002.

Noë, Alva. *Out of Our Heads: Why You Are Not Your Brain, and Other Lessons from the Biology of Consciousness*. New York: Hill and Wang, 2009.

Nordal, Katherine C. "Where Has All the Psychotherapy Gone?" *APA Monitor* 41, no. 10 (2010): 17.

Olfson, Mark, Carlos Blanco, Shuai Wang, Gonzalo Laje, and Christoph U. Correll. "National Trends in the Mental Health Care of Children, Adolescents, and Adults by Office-Based Physicians." *JAMA Psychiatry* 71, no. 1 (2014): 81–90.

Olfson, Mark, and Gerald L. Klerman. "Trends in the Prescription of Psychotropic Medications: The Role of Physician Specialty." *Medical Care* 31 (1993): 559–64.

Olfson, Mark, and Steven C. Marcus. "National Trends in Outpatient Psychotherapy." *American Journal of Psychiatry* 167 (2010): 1456–63.

Palumbo, Francis B., and C. Daniel Mullins, "The Development of Direct-to-Consumer Prescription Drug Advertising Regulation." *Food and Drug Law Journal* 57 (2002): 423–43.

Parry, Hugh J. "Use of Psychotropic Drugs by U.S. Adults." *Public Health Reports* 83 (1968): 799–810.

Parry, Hugh J., Mitchell B. Balter, Glen D. Mellinger, Ira H. Cisin, and Dean I. Manheimer. "National Patterns of Psychotherapeutic Drug Use." *Archives of General Psychiatry* 28 (1973): 769–83.

Parsons, Talcott. *The Social System*. Glencoe, IL: Free Press, 1951.

Payer, Lynn. *Disease-Mongers: How Doctors, Drug Companies, and Insurers Are Making You Feel Sick*. New York: John Wiley & Sons, 1992.

Payton, Andrew R., and Peggy A. Thoits. "Medicalization, Direct-to-Consumer Advertising, and Mental Illness Stigma." *Society and Mental Health* 1 (2011): 55–70.

Peale, Norman Vincent. *The Power of Positive Thinking*. New York, Prentice-Hall, 1952.

Pescosolido, Bernice A., Jack K. Martin, J. Scott Long, Tait R. Medina, Jo C. Phelan, and Bruce G. Link. "'A Disease Like Any Other'? A Decade of Change in Public Reactions to Schizophrenia, Depression, and Alcohol Dependence." *American Journal of Psychiatry* 167 (2010): 1321–30.

Pieters, Toine, and Stephen Snelders. "Psychotropic Drug Use: Between Healing and Enhancing the Mind." *Neuroethics* 2 (2009): 63–73.

Pitts-Taylor, Victoria. "The Plastic Brain: Neoliberalism and the Neuronal Self." *Health* 14 (2010): 635–52.

Pratt, Laura A., Debra J. Brody, and Qiuping Gu. "Antidepressant Use in Persons Aged 12 and Over: United States, 2005–2008." *NCHS Data Brief*, no. 76 (October 2011).

Prins, Marijn A., Peter F. M. Verhaak, Jozien M. Bensing, and Klaas van der Meer. "Health Beliefs and Perceived Need for Mental Health Care of Anxiety and Depression—the Patients' Perspective Explored." *Clinical Psychology Review* 28 (2008): 1038–58.

Rahav, Michael. "Public Images of the Mentally Ill in Israel." *International Journal of Mental Health* 15, no. 4 (1987): 59–69.

Read, John. "Why Promoting Biological Ideology Increases Prejudice against People Labelled 'Schizophrenic.'" *Australian Psychologist* 42 (2007): 118–28.

Read, John, and Jacqui Dillon. *Models of Madness*. 2nd ed. New York: Routledge, 2013.

Reed, Isaac. *Interpretation and Social Knowledge*. Chicago: University of Chicago Press, 2011.

Reynolds, Charles F., David A. Lewis, Thomas Detre, Alan F. Schatzberg, and David J. Kupfer. "The Future of Psychiatry as Clinical Neuroscience." *Academic Medicine* 84 (2009): 446–50.

Rieff, Philip. *Freud: The Mind of the Moralist*. Chicago: University of Chicago Press, 1959.

Rieff, Philip. *The Triumph of the Therapeutic: Uses of Faith after Freud*. New York: Harper and Row, 1966.

Riesman, David. *The Lonely Crowd: A Study of the Changing American Character*. New Haven, CT: Yale University Press, 1950.

Rittmannsberger, H. "The Use of Drug Monotherapy in Psychiatric Inpatient Treatment." *Progress in Neuro-Psychopharmacology & Biological Psychiatry* 26 (2002): 547–51.

Roberts, Robert C. "Emotions and Culture." In *The Emotions and Cultural Analysis*, edited by Ana Marta González, 19–30. Burlington, VT: Ashgate, 2012.

Rogers, Carl. *On Becoming a Person*. Boston: Houghton Mifflin, 1961.

Rosa, Hartmut. *Social Acceleration: A New Theory of Modernity*. New York: Columbia University Press, 2013.

Rosaldo, Renato. *Culture and Truth: The Remaking of Social Analysis*. London: Routledge, 1989.

Rose, Nikolas. *Inventing Our Selves: Psychology, Power, and Personhood*. Cambridge: Cambridge University Press, 1998.

Rose, Nikolas. *The Politics of Life Itself: Biomedicine, Power, and Subjectivity in the Twenty-First Century*. Princeton, NJ: Princeton University Press, 2007.

Rose, Nikolas, and Carlos Novas. "Biological Citizenship." In *Global Assemblages: Technology, Politics, and Ethics as Anthropological Problems*, edited by Aihwa Ong and Stephen J. Collier, 439–63. Malden, MA: Blackwell, 2005.

Rosen, R. D. *Psychobabble: Fast Talk and Quick Cure in the Era of Feeling*. New York: Atheneum, 1977.

Rosenberg, Alex. "Why You Don't Know Your Own Mind." *New York Times*, July 18, 2016.

Rosenberg, Charles E. *Our Present Complaint: American Medicine, Then and Now*. Baltimore, MD: Johns Hopkins University Press, 2007.

Rosenhan, D. L. "On Being Sane in Insane Places." *Science* 179 (1973): 250–58.

Ross, David A., Michael J. Travis, and Melissa R. Arbuckle. "The Future of Psychiatry as Clinical Neuroscience: Why Not Now?" *JAMA Psychiatry* 72 (2015): 413–14.

Ross, Jerilyn. *Triumph over Fear*. New York: Bantam Books, 1994.

Rundell, John. "Durkheim and the Reflexive Condition of Modernity." In *Recognition*,

*Work, Politics: New Directions in French Critical Theory*, edited by Jean-Philippe De-ranty, Danielle Petherbridge, John Rundell, and Robert Sinnerbrink, 203–30. Leiden: Brill, 2007.

Rüsch, Nicolas, Andrew R. Todd, Galen V. Bodenhausen, and Patrick W. Corrigan. "Bio-genetic Models of Psychopathology, Implicit Guilt, and Mental Illness Stigma." *Psychiatry Research* 179 (2010): 328–32.

Sadowsky, Jonathan. "Beyond the Metaphor of the Pendulum: Electroconvulsive Ther-apy, Psychoanalysis, and the Styles of American Psychiatry." *Journal of the History of Medicine and Allied Sciences* 61 (2005): 1–25.

Sandel, Michael J. *Liberalism and the Limits of Justice*. Cambridge: Cambridge University Press, 1982.

Sandomir, Richard. "Spencer Johnson, Author of Pithy Best-Sellers, Dies at 78." *New York Times*, July 8, 2017, D6.

Sarbin, Theodore R., and James C. Mancuso. "Failure of a Moral Enterprise: Attitudes of the Public toward Mental Illness." *Journal of Consulting and Clinical Psychology* 35 (1970): 159–73.

Sayer, Andrew. *Why Things Matter to People: Social Science, Values and Ethical Life*. Cam-bridge: Cambridge University Press, 2011.

Scambler, Graham, and Anthony Hopkins. "Being Epileptic: Coming to Terms with Stigma." *Sociology of Health & Illness* 8 (1986): 26–43.

Schildkraut, Joseph J., and Seymour S. Kety. "Biogenic Amines and Emotion." *Science* 156 (1967): 21–30.

Schiraldi, Glenn R. *The Self-Esteem Workbook*. 2nd ed. Oakland, CA: New Harbinger Pub-lications, 2016.

Schneider, Joseph W., and Peter Conrad. "In the Closet with Illness: Epilepsy, Stigma Po-tential and Information Control." *Social Problems* 28 (1980): 32–44.

Schneier, Franklin, and Lawrence Welkowitz, *The Hidden Face of Shyness*. New York: Avon Books, 1996.

Schomerus, G., C. Schwahn, A. Holzinger, P. W. Corrigan, H. J. Grabe, M. G. Carta, and M. C. Angermeyer. "Evolution of Public Attitudes about Mental Illness: A Systematic Review and Meta-Analysis." *Acta Psychiatrica Scandinavica* 125 (2012): 440–52.

Schreiber, Rita, and Gwen Hartrick. "Keeping It Together: How Women Use the Biomedi-cal Explanatory Model to Manage the Stigma of Depression." *Issues in Mental Health Nursing* 23 (2002): 91–105.

Schwartz, Barry. "The Tyranny of Choice." *Scientific American* 290 (April 2004): 70–75.

Scull, Andrew. *Madness in Civilization*. Princeton, NJ: Princeton University Press, 2015.

Sennett, Richard. *The Corrosion of Character: The Personal Consequences of Work in the New Capitalism*. New York: W. W. Norton, 1998.

Sennett, Richard. *The Culture of the New Capitalism*. New Haven, CT: Yale University Press, 2006.

Shapiro, Sam, and Seymour H. Baron. "Prescriptions for Psychotropic Drugs in a Non-institutional Population." *Public Health Reports* 76, no. 6 (1961): 481–88.

Siegel, Lyn. "DTC Advertising: Bane . . . or Blessing? A 360-degree View." *Pharmaceutical Executive* 20, no. 10 (October 2000): 140–52.

Simko, Christina. "The Problem of Suffering in the Age of Prozac: A Case Study of the Depression Memoir." In *To Fix or to Heal: Patient Care, Public Health, and the Limits of Biomedicine*, edited by Joseph E. Davis and Ana Marta González, 63–83. New York: New York University Press, 2016.

Slater, Lauren. *Prozac Diary*. New York: Penguin Books, 1999.

Smith, Christian. *Moral, Believing Animals: Human Personhood and Culture*. New York: Oxford University Press, 2003.

Smith, Mickey C. *A Social History of the Minor Tranquilizers: The Quest for Small Comfort in the Age of Anxiety*. Binghamton, NY: Pharmaceutical Products Press, 1991.

Snyder, Solomon H., Shailesh P. Banerjee, Henry I. Yamamura, and David Greenberg. "Drugs, Neurotransmitters, and Schizophrenia." *Science* 184 (1974): 1243–53.

Solomon, Andrew. *The Noonday Demon: An Atlas of Depression*. New York: Scribner, 2001.

Spaemann, Robert. *Persons: The Difference between "Someone" and "Something."* New York: Oxford University Press, 2007.

Spitzer, Robert L., Janet B. W. Williams, and Andrew E. Skodol. "DSM-III: The Major Achievements and an Overview." *American Journal of Psychiatry* 137 (1980): 151–64.

Srole, Leo, Thomas S. Langner, Stanley T. Michael, Marvin K. Opler, and Thomas A. C. Rennie. *Mental Health in the Metropolis: The Midtown Manhattan Study*. New York: McGraw-Hill, 1962.

Stearns, Peter N. *American Cool: Constructing a Twentieth-Century Emotional Style*. New York: New York University Press, 1994.

Stein, Murray B., and John R. Walker. *Triumph over Shyness*. New York: McGraw-Hill, 2002.

Stepnisky, Jeffrey N. "Narrative Magic and the Construction of Selfhood in Antidepressant Advertising." *Bulletin of Science, Technology & Society* 27 (2007): 24–36.

Stossel, Scott. *My Age of Anxiety*. New York: Knopf, 2014.

Stromberg, Peter G. "The Impression Point: Synthesis of Symbol and Self." *Ethos* 13 (1985): 56–74.

Stuart, Heather, and Julio Arboleda-Florez. "Community Attitudes toward People with Schizophrenia." *Canadian Journal of Psychiatry* 46 (2001): 245–52.

Sullivan, Helen W., Kathryn J. Aikin, Eunice Chung-Davies, and Michael Wade. "Prescription Drug Promotion from 2001–2014: Data from the U.S. Food and Drug Administration." *PLoS One* 11, no. 5 (2016): e0155035.

Taylor, Charles. *The Ethics of Authenticity*. Cambridge, MA: Harvard University Press, 1991.

Taylor, Charles. *Human Agency and Language*. Cambridge: Cambridge University Press, 1985.

Taylor, Charles. *Modern Social Imaginaries*. Durham, NC: Duke University Press, 2004.

Taylor, Charles. *Sources of the Self: The Making of the Modern Identity*. Cambridge, MA: Harvard University Press, 1989.

Thomson, Irene Taviss. *In Conflict No Longer: Self and Society in Contemporary America*. Lanham, MD: Rowman & Littlefield, 2000.

Tone, Andrea. *The Age of Anxiety: A History of America's Turbulent Affair with Tranquilizers*. New York: Basic Books, 2009.

Totton, Nick. "Two Ways of Being Helpful." *CPJ: Counselling & Psychotherapy Journal* 15, no. 10 (2004): 5–8.

Tourney, Garfield. "History of Biological Psychiatry in America." *American Journal of Psychiatry* 126 (1969): 29–42.

Trethowan, W. H. "Pills for Personal Problems." *British Medical Journal* 3 (1975): 749–51.

Trilling, Lionel. *Sincerity and Authenticity*. New York: Harcourt Brace Jovanovich, 1980. First published 1972.

Turner, Victor. *The Ritual Process: Structure and Anti-Structure*. New York: Aldine de Gruyter 1995. First published 1969.

US Department of Health and Human Services. *Mental Health: A Report of the Surgeon General*. Rockville, MD: US Department of Health and Human Services, 1999.

Van Voorhees, Benjamin W., Joshua Fogel, Thomas K. Houston, Lisa A. Cooper, Nae-Yuh

Wang, and Daniel E. Ford. "Beliefs and Attitudes Associated with the Intention to Not Accept the Diagnosis of Depression among Young Adults." *Annals of Family Medicine* 3 (2005): 38–46.

Vaughan, Susan C. *The Talking Cure: The Science behind Psychotherapy*. New York: Grosset/ Putnam, 1997.

Ventola, C. Lee. "Direct-to-Consumer Pharmaceutical Advertising: Therapeutic or Toxic?" *Pharmacy and Therapeutics* 36 (2011): 669–74, 681–84.

Veroff, Joseph, Elizabeth Douvan, and Richard A. Kulka. *The Inner American: A Self-Portrait from 1957 to 1976*. New York: Basic Books, 1981.

Veroff, Joseph, Richard A. Kulka, and Elizabeth Douvan. *Mental Health in America: Patterns of Help-Seeking from 1957 to 1976*. New York: Basic Books, 1981.

Visser, Susanna N., Melissa L. Danielson, Rebecca H. Bitsko, Joseph R. Holbrook, Michael D. Kogan, Reem M. Ghandour, Ruth Perou, and Stephen J. Blumberg. "Trends in the Parent-Report of Health Care Provider-Diagnosed and Medicated Attention-Deficit/Hyperactivity Disorder: United States, 2003–2011." *Journal of the American Academy of Child and Adolescent Psychiatry* 53 (2014): 34–46.

Vrecko, Scott. "Just How Cognitive Is 'Cognitive Enhancement'? On the Significance of Emotions in University Students' Experiences with Study Drugs." *AJOB Neuroscience* 4 (2013): 4–12.

Waldron, Ingrid. "Increased Prescribing of Valium, Librium, and Other Drugs—an Example of the Influence of Economic and Social Factors on the Practice of Medicine." *International Journal of Health Services* 7 (1977): 37–62.

Wall, Terri N., and Jeffrey A. Hayes. "Depressed Clients' Attributions of Responsibility for the Causes of and Solutions to Their Problems." *Journal of Counseling and Development* 78 (2000): 81–86.

Waschbusch, Daniel A., William E. Pelham, James Waxmonsky, and Charlotte Johnston. "Are There Placebo Effects in the Medication Treatment of Children with Attention-Deficit Hyperactivity Disorder?" *Journal of Developmental and Behavioral Pediatrics* 30 (2009): 158–68.

Weber, Max. "'Objectivity' in Social Science and Social Policy." In *The Methodology of the Social Sciences*, edited by Edward A. Shils and Henry A. Finch, 49–112. New York: Free Press, 1949. First published 1904.

Weber, Max. "The Social Psychology of the World Religions." In *From Max Weber: Essays in Sociology*, edited by H. H. Gerth and C. Wright Mills, 267–301. New York: Oxford University Press, 1958. First published 1946.

Weber, Max. *The Sociology of Religion*. Boston: Beacon Press, 1963. First published 1922.

Weinstein, Raymond M., and Norman Q. Brill. "Social Class and Patients' Perceptions of Mental Illness." *Psychiatric Quarterly* 45 (1971): 35–44.

Weiss, Gabrielle, John Werry, Klaus Minde, Virginia Douglas, and Donald Sykes. "Studies on the Hyperactive Child—V: The Effects of Dextroamphetamine and Chlorpromazine on Behaviour and Intellectual Functioning." *Journal of Child Psychology and Psychiatry* 9 (1968): 145–56.

Whyte, Susan Reynolds, Sjaak van der Geest, and Anita Hardon. *Social Lives of Medicines*. Cambridge: Cambridge University Press, 2003.

Wilkinson, Iain. *Suffering: A Sociological Introduction*. Malden, MA: Polity, 2005.

Williams, Simon J., and Michael Calnan. "The 'Limits' of Medicalization? Modern Medicine and the Lay Populace in 'Late' Modernity." *Social Science and Medicine* 42 (1996): 1609–20.

Williamson, Judith. *Decoding Advertisements*. London: Marion Boyars, 1978.

Wilson, Edward O. *The Meaning of Human Existence*. New York: Liveright, 2014.

Wilson, Mitchell. "DSM-III and the Transformation of American Psychiatry: A History." *American Journal of Psychiatry* 150 (1993): 399–410.

Wilson, Timothy D. *Strangers to Ourselves: Discovering the Adaptive Unconscious*. Cambridge, MA: Harvard University Press, 2002.

Winkelman, N. William, Jr. "Chlorpromazine in the Treatment of Neuropsychiatric Disorders." *Journal of the American Medical Association* 155, no. 1 (1954): 18–21.

Wuthnow, Robert, James Davison Hunter, Albert Bergesen, and Edith Kurzweil. *Cultural Analysis: The Work of Peter L. Berger, Mary Douglas, Michel Foucault, and Jurgen Habermas*. Boston: Routledge & Kegan Paul, 1984.

Wynne, Brian. "Knowledges in Context." *Science, Technology, and Human Values* 16, no. 1 (1991): 111–21.

Zimmerman, Mark, and Janine Galione. "Psychiatrists' and Nonpsychiatrist Physicians' Reported Use of the *DSM-IV* Criteria for Major Depressive Disorder." *Journal of Clinical Psychiatry* 71 (2010): 235–38.

Zito, Julie Magno. "Pharmacoepidemiology: Recent Findings and Challenges for Child and Adolescent Psychopharmacology." *Journal of Clinical Psychiatry* 68 (2007): 966–67.